Control Systems: A Historical and Philosophical Perspective

Ali Khaki-Sedigh

Control Systems: A Historical and Philosophical Perspective

History and Philosophy of Control Systems

Springer

Ali Khaki-Sedigh
Department of Electrical Engineering
K. N. Toosi University of Technology
Tehran, Iran

ISBN 978-981-96-7425-1 ISBN 978-981-96-7426-8 (eBook)
https://doi.org/10.1007/978-981-96-7426-8

This Springer imprint is published by the registered company Springer Nature Singapore Pte Ltd.
The registered company address is: 152 Beach Road, #21-01/04 Gateway East, Singapore 189721, Singapore

Preface

An introductory note. This book is a culmination of my interest in control engineering, history, and philosophy, stemming from my experiences as a master's student in the Control Systems Centre at the University of Manchester Institute of Science and Technology (UMIST), which later merged with the Victoria University of Manchester. I was particularly fascinated by the seminars conducted by the late Professor H. H. Rosenbrock, who emphasised the philosophy of engineering and ethical issues. These discussions were both new and captivating, shifting our focus from detailed mathematical and technical discussions to broader considerations that opened new horizons in our minds. That experience ignited my curiosity about how historical context and philosophical insights have influenced control engineering practices. Throughout my career, I have often noticed that discussions among control engineers tend to focus narrowly on contemporary challenges, while the rich lessons from history and philosophy are frequently overlooked. I believe that integrating these fields can enhance our understanding and lead to more innovative engineering solutions. Motivated by this realisation, I explored the historical dimensions of control engineering in a book published in Persian by K. N. Toosi University of Technology several years ago. In this monograph, I aim to add a historical-philosophical dimension to control systems studies, hoping to inspire fellow engineers and students to consider a broader perspective.

Historical-philosophical studies and engineering. Many engineering departments, including academic staff and students, are largely unaware of their fields' historical and philosophical developments, often functioning as isolated entities with limited intellectual interaction and a restricted understanding of broader concepts. However, with the introduction of ever-increasing complex engineering problems, engineers face a wide range of conflicting technical issues with no obvious solution. In such cases, engineers are often involved in multiple disciplines. For professionals engaged in engineering practice, studying analytical history and applied philosophy can significantly enhance critical thinking, soft skills, and the ability to tackle multidimensional engineering problems. These disciplines cultivate essential problem-solving and decision-making skills, which are invaluable in engineering. Philosophical studies can provide a deep and foundational understanding of the underlying engineering design processes and ethical principles, which are crucial for engineers facing ethical dilemmas in their professional lives.

History and philosophy of control systems analysis and design. Although revisiting the historical development of control systems analysis and design may not directly contribute to solving today's specific control engineering problems, it can provide valuable insight into the necessity of a holistic perspective and a deeper understanding of the challenges faced by control systems engineers. This is achieved by incorporating critical views of various theories, highlighting the need for paradigm shifts in control systems, and connecting historical discussions with the philosophical aspects of design. This book examines the philosophical foundations of control systems, focusing on cybernetics and systems theory, which have shaped control systems analysis and design. Exploring these foundations helps clarify their conceptual basis and broader implications.

Additionally, philosophical studies can inspire innovation by showing how past scientists and engineers solved challenges using philosophical concepts, offering new ideas and approaches to contemporary problems. Control systems are closely linked to multiple engineering and scientific fields. Adopting a historical and philosophical perspective encourages an interdisciplinary approach, leading to a more comprehensive understanding of control systems. Moreover, viewing control systems through a philosophical lens can highlight ethical issues surrounding the use of technology in control and automation, particularly as autonomous and intelligent systems continue to develop. Hence, incorporating historical and philosophical perspectives into control systems education enhances the learning experience by providing students with a broader context for theory and practice. This approach fosters a well-rounded understanding of the subject, encouraging students to consider technical aspects and societal, cultural, and ethical dimensions.

Aims and objectives. This book serves as an introduction to the historical and philosophical perspectives of control systems. It offers an overview of key historical developments in control theory, from antiquity to the major achievements of the twentieth century. Additionally, it presents philosophical perspectives on control theory, connecting historical developments with philosophical concepts. Drawing on ideas from the philosophy of science and historical discussions, the book explores paradigm shifts in control systems engineering. Aimed at students and researchers interested in the history and philosophy of control systems, it provides a fresh perspective on the field's evolution. The book presents foundational material, offering readers a general overview without delving into mathematical discussions, thus broadening its appeal to those in the history and philosophy of engineering. Given the limited number of published works specialising in the history and philosophy of control systems, the references cited in this book form the core resources for its content.

Overview of the book. The book is organised as follows. Chapter 1 introduces the connections and interrelations between history, philosophy, and engineering. It then presents the concepts of feedback and control systems design and emphasises the significance of the historical and philosophical perspectives in studying feedback control systems. The chapter continues with general remarks on engineering ontology to distinguish the interconnected and divergent concepts and methods of engineering and science. Finally, a general systems classification and a note on complexity and complex adaptive systems are provided.

Chapter 2 is the first of three chapters on historical perspectives. It provides a concise history of control systems from antiquity to the pre-classical control era. The primitive control designs up to about 1200 AD are briefly reviewed, and then the developments from the pre-classical control era leading to the major achievements of the twentieth century are outlined.

Control systems from classical to modern control are presented in Chap. 3. Key points, including the foundations of classical control systems and the effects of World War II on them, are briefly discussed. Then, the emergence of mathematical system theory, particularly the state-space-based mathematical control theory period, is introduced. Poincaré's impact and Kalman's legacy are outlined, with the evolution of filtering techniques highlighted as a key change from frequency response to state-space theory. Finally, the Achilles' heel of mathematical system theory is discussed concerning the paradigm shift in control theory.

Chapter 4 is the last chapter on the history of control systems and encompasses the postmodern control systems period. The historical developments of the three main methodologies—robust control, adaptive control, and intelligent control—are discussed in this period. It is argued that robustness, adaptation, and intelligence are the central tenets of the postmodern control paradigm. It is important to note that the chapter does not examine specific control design strategies within each of these methodologies but reviews them as different schools of thought with various design philosophies.

Models are the most widely used tools in many engineering frameworks. Chapter 5 focuses on models and harnessing them for control systems analysis and design. Different approaches for model derivation include analytical mathematical modelling, modelling from data: conventional system identification, and data-driven modelling. Some philosophical considerations are noted on the subject. These are categorised as philosophical foundations of modelling, ontological discussions, the model generation process, and the inductive-deductive approach to modelling. Then, the Certainty Equivalence Principle for feedback control is discussed. Model uncertainty is of utmost importance when applying models in control engineering. Questions such as model uncertainty types, sources of model uncertainty, and epistemological aspects of model uncertainty are addressed in this chapter. Finally, the trio of uncertainty combat: robustness, adaptation, and intelligence are discussed for uncertainty mitigation strategies.

Chapter 6 explores the advent of system theory and its impact on control systems design. First, a philosophical perspective on system theory is presented. The general system theory is then discussed from control systems design, cybernetics, and modelling perspectives. Mathematical system theory is identified as a key achievement of system theory in the last century, with early developments and the concepts of state space being discussed. The severe limitations of mathematical system theory, including assumptions and simplifications, causality and feedback, uncertainty and transparency, system characteristics versus model characteristics, and the important issue of the theory-practice gap, are also explored.

Popper's philosophy of falsification has been introduced into control systems analysis and design. Chapter 7 discusses the basic concepts of engineering knowledge and falsification in an attempt to adapt this philosophical framework to failed designs. The fundamental issues of the characteristics of a scientific theory, the concept of theory-laden observations, and the problem of infinite regressors are presented. Then, falsification is introduced as an alternative to induction-based feedback control philosophy. Finally, the application of falsification in actual control engineering practice is briefly presented.

Chapter 8 presents the paradigm shifts in control theory from Kuhn's revolutionary perspective. The basic concepts and definitions of scientific revolutions and paradigm shifts are introduced. These ideas are then applied to the context of control systems design from antiquity to the present day, illustrating the revolutions in control systems design through Kuhn's framework. Four main paradigm shifts in control systems design are identified and discussed.

The issue of postmodernism in control systems design is discussed in Chap. 9. The chapter describes the transition from modernism to postmodernism in control systems design methodologies and provides a classification of modernist and postmodernist control theories. Subsequently, the implications of postmodernism, pluralism, and internal realism in control systems design are presented.

Chapter 10 investigates the challenges and future trends in control systems proposed by experts in the field from the 1960s to the present. It presents the early 1963 Fuller paper on research directions in control, leading up to the most recent 2030 control systems roadmap. Finally, the important issue of ethical considerations in control engineering is discussed.

To the best of the author's knowledge, this book is the first published monograph on control systems from a historical-philosophical perspective; however, it may have shortcomings, and many issues discussed are subject to ongoing technical debates. The author welcomes experts in the field to share their views and insights to enrich the discourse further.

Tehran, Iran
May 2025

Ali Khaki-Sedigh

Acknowledgements

The creation of this book has been profoundly supported by the generous contributions of many postgraduate students and colleagues, who devoted their time to offering expert insights, constructive feedback, and helpful corrections throughout the process. I am deeply grateful to **Seyyed Hojjatolhagh Hosseini** for encouraging me to write this book and to the individuals listed below, whose dedicated efforts have been instrumental in enhancing the quality and content of this work:

- **Hossein Sheikh Rezaee** from the Iranian Institute of Philosophy for his insightful discussions, the introduction of numerous valuable references, and his instrumental comments, which notably enhanced various parts of the book.
- **Mohammad Ataei** from the University of Isfahan for his valuable comments on the book and for reviewing the manuscript.
- **Abdolah Shamisa** from Khatam University for his support and valuable general comments on the book.
- **Ahmad Reza Tahsiri** from K. N. Toosi University of Technology for his valuable comments on the early chapters.
- **Hamed Kebriaei** from the University of Tehran for his discussions and valuable comments on Chap. 10.
- **Sadegh Vaez-Zadeh** from the University of Tehran for his encouragement and valuable comments.
- **Mahdi Aliyari-Shoorehdeli** from K. N. Toosi University of Technology for his useful discussions on the artificial intelligence and soft computing sections.
- **Sepideh Nasrollahi**, My Ph.D. student from the Electrical Engineering Department at K. N. Toosi University of Technology, for her proofreading and excellent illustrations of the figures in the book.
- The two anonymous reviewers of my book *An Introduction to Data-Driven Control Systems*, published in 2023 by Wiley-IEEE Press, for their encouraging and inspiring comments on Chap. 1, particularly on the philosophical discussions that led to the development of this present book.

- Gratitude is extended to **ChatGPT** and **DeepSeek** for their assistance in discussing and clarifying ideas and providing insightful text consultation throughout the development of this book.

Finally, I would like to express my sincere appreciation to **Khatam University** for its financial support of the project on philosophy and engineering. This assistance has enabled the organisation of a one-day seminar on *Philosophy and Engineering*, held on 30th November 2023, as well as an autumn school on *Artificial Intelligence: Philosophical and Engineering Perspectives*, held on 16–17 October 2024, both at **Khatam University**. Moreover, this support has also facilitated the publication of this book. Without this backing, the realisation of this publication would not have been feasible, and I am deeply grateful for **Khatam University**'s commitment to fostering scholarly work in the fields of philosophy and engineering. The collaboration between the Departments of Philosophy and Electrical Engineering at **Khatam University** has been a unique experience and has paved the way for future partnerships between philosophers and engineers.

Last but not least, I am deeply grateful to my family for their unwavering support and understanding, which allowed me to devote much of my holidays, weekends, and evenings to completing this book.

Contents

Acronyms

ABET	Accreditation Board for Engineering and Technology
ADALINE	ADAptive LINear Element
AI	Artificial Intelligence
AT&T	American Telephone and Telegraph Company
CAS	Complex Adaptive Systems
CPHS	Cyber-Physical Human Systems
CPS	Cyber-Physical Systems
DL	Dynamic Linearisation
EA	Evolutionary Algorithms
ECPD	Engineering Council for Professional Development
ETH	Eidgenössische Technische Hochschule Zürich
FOPDT	First Order Plus Dead Time
GA	Genetic Algorithms
GP	Genetic Programming
GST	General Systems Theory
ICI	Imperial Chemical Industries
ICM	International Congress of Mathematicians
IFAC	International Federation of Automatic Control
IMU	International Mathematical Union
LQ	Linear Quadratic
LQG	Linear-Quadratic-Gaussian
LQG/LTR	Linear-Quadratic-Gaussian/Loop Transfer Recovery
MADALINE	Multiple ADAptive LINear Element
MIMO	Multi Input Multi Output
MIT	Massachusetts Institute of Technology
MPC	Model Predictive Control
MPUM	Most Powerful Unfalsified Model
MRAC	Model Reference Adaptive Control
MST	Mathematical System Theory
NDRC	National Defense Research Committee
OSRD	Office of Scientific Research and Development

PID	Proportional-Integral-Derivative
PPD	Pseudo-Partial Derivative
PSO	Particle Swarm Optimisation
PWM	Pulse-Width Modulation
QFT	Quantitative Feedback Theory
RAND	Adopted from a contraction of the term research and development
RGA	Relative Gain Array
RLS	Recursive Least Squares
SISO	Single Input Single Output
SOC	Second-Order Cybernetics
SoS	Systems of Systems
SSV	Structured Singular Value
STEM	Science, Technology, Engineering and Mathematics
SVD	Singular Value Decomposition
UASC	Unfalsified Adaptive Switching Control
VDE	Verband der Elektrotechnik
VDI	Verein Deutscher Ingenieure
VFD	Variable Frequency Drive

Chapter 1
Introduction

Engineers of the world philosophize! You have nothing to lose but your silence!
—Carl Mitcham

1.1 History, Philosophy, and Engineering

In the dynamic, ever-evolving field of engineering education, guiding principles are essential to enhance program quality and relevance in response to a rapidly changing environment. ABET[1] provides specific criteria for all accredited engineering programs [1]. These criteria aim to improve engineering education systematically, ensuring it meets the needs of its stakeholders. ABET's third criterion is *student outcomes*, where the graduate of an engineering programme becomes prepared to enter the engineering profession by attaining these outcomes. The first of the seven outcomes is the ability to identify, formulate, and solve *complex engineering problems*. Also, four other outcomes are directly or indirectly related to a graduate's *soft skills*.

Complex engineering problems are defined by ABET as problems that include one or more of the following characteristics: involving wide-ranging or conflicting technical issues, having no obvious solution, addressing problems not encompassed by current standards and codes, involving diverse groups of stakeholders, including many parts or sub-problems, involving multiple disciplines, or having significant consequences in a range of contexts.

For engineering students and professional engineers, studying analytical history and applied philosophy can significantly enhance their critical thinking and soft skills and handling of multidimensional engineering problems. Philosophical studies can provide a deep and foundational understanding of the underlying engineering design processes and also of ethical principles, crucial for engineers facing ethical dilemmas in their professional lives.

[1] Accreditation Board for Engineering and Technology.

A. Khaki-Sedigh, *Control Systems: A Historical and Philosophical Perspective*,
https://doi.org/10.1007/978-981-96-7426-8_1

Additionally, history offers a perspective on technological advancements and engineering practices, aiding students in understanding the evolution and societal impact of engineering.

Furthermore, considering philosophical and historical features often emphasises effective communication—an essential skill for engineers collaborating in multidisciplinary teams or interacting with non-engineers. This emphasis on communication can help bridge the gap between technical and non-technical problem-solving aspects. The deficiency of a connection between social sciences and STEM[2] disciplines, notably in engineering, has sparked numerous discussions. Many argue for integrating social sciences to enhance creativity, responsiveness, and a better understanding of the human condition among scientists and engineers.

However, in engineering research and education, there is a noticeable absence of a link between technical studies and historical/philosophical aspects. Nevertheless, by engaging engineering students and researchers with their field's historical and philosophical dimensions, we may find that their interests and insights deepen.

Moreover, engineers should acknowledge and embrace their role in a broader historical and educational context and recognise that they contribute to a grand tradition of learning, encompassing both success and failure. This recognition allows engineers to achieve goals such as preserving valuable knowledge, drawing inspiration from past innovations, accessing diverse resources, and promoting deeper engagement with the subject matter, thereby overcoming the initial challenges associated with these objectives [2].

1.1.1 History and Engineering

The history of engineering, when coupled with an inclusive description and interpretation of changes, can play a crucial role in reshaping the perspectives of engineers and engineering students. This transformation of perspectives involves moving away from a narrow emphasis on complicated technical solutions and towards a more comprehensive perspective in the context of problem-solving.

While there is a widespread belief in the importance of *learning the lessons of the past*, it is essential to recognise that engineering events are continually evolving, its history never repeats itself precisely, and its predictive potential is, therefore, limited. Although it can shed light on potential preconditions leading to a specific outcome and draw attention to the influential characteristics of actors or contexts shaping events, it cannot serve as an infallible guide for contemporary behaviour.

Conversely, looking at how things are now compared to how they were before can help us see what parts of our current knowledge and professionalism are temporary and what parts will last. This contributes significantly to our understanding of the present. The history of engineering also serves as "an inventory of alternatives" or,

[2] Science, Technology, Engineering and Mathematics.

in the words of a cultural historian, as a means to showcase that "things do not have to be the way they are now," and they could be different with other alternatives.

However, it is important to note that historical narratives may not comprehensively describe alternative solutions to a particular *engineering* problem. Instead, their primary utility for engineering education lies in their demonstration of why a holistic view and an understanding of the challenges faced by engineers must be inherent to engineering practice [3].

The approach in this book involves incorporating critics' views of various theories to demonstrate the necessity of paradigm shifts in control systems. It aims to connect historical discussions with the philosophical aspects of designs.

Finally, it is important to mention that engineering history can be primarily about the engineering profession and its evolution, or it could consider its technical realisations, such as infrastructures or machines. Also, the social and environmental dimensions of engineering history are important and well-studied. However, this book will exclusively address the technical aspects. For a perspective on sustainability and social justice in engineering practice, refer to [3, 4] and the references therein.

1.1.2 Philosophy and Engineering

The philosophy of engineering is shaped and informed by two longer established significant branches of philosophical inquiry—namely, the *philosophy of science* and the *philosophy of technology*. The philosophy of science deals with the fundamental principles, methodologies, and implications of scientific practices, while the philosophy of technology explores technology's nature, development, and impact. Despite these ties, the philosophy of engineering aims to establish itself as a unique and independent field of study [5]. This autonomy is driven by a set of specific goals, both philosophical and practical.

To clarify the subject, we first distinguish between the philosophy of engineering and engineering philosophy. The delicate difference lies in the fact that engineering philosophy represents the mindset or approach engineers adopt in their daily tasks, which may or may not involve a formal philosophical framework. On the other hand, the philosophy of engineering, at its core, is committed to reflection, using philosophical perspectives to investigate engineering practice. As a result, it can be employed to shape or enlighten engineering philosophy [6].

For example, in the design of multivariable controllers for a plant with multi-inputs and multi-outputs, two control engineering design philosophies could be the *decentralised* and the *centralised* approaches. Also, in [7], new power plant control philosophies are discussed, which are design strategies dependent upon reliability, operability, maintainability, and economic factors. On the other hand, when we discuss the role of induction, or falsification, and process philosophy underlying general system theory and its consequences, the respective formal philosophical perspectives and their implications in control design are of concern.

In [8], the significance of philosophy in engineering is thoroughly examined. It is stated that "When engineers and engineering students dismiss philosophical analysis and reflection as marginal to the practice of engineering, they are mistaken on at least two counts: historical and professional," and engineers are coined as "the unacknowledged philosophers of the postmodern world." The reasons given for the importance of philosophy in engineering are:

- To better define, acknowledge, and comprehend the mission of engineering, utilising philosophical principles and arguments.
- To address the professional problems associated with engineering practice that cannot be resolved by mere engineering methods. These issues include engineering ethics, the social responsibility of engineers, environmental concerns, safety, risk, and values in engineering practice.
- Directed philosophical studies are crucial for engineering education to promote critical thinking and handling complex, multidimensional problems.
- Engineering has a philosophical dimension, and by engaging with philosophical principles, engineers can reflect on their own understanding, practices, and perspectives. Hence, they can gain deeper insight and understanding by self-examination and thoughtful consideration of their engineering designs.

Engineering draws upon knowledge and inspiration from various fields relevant to problem-solving. An engineer generally utilises the principles and techniques of basic sciences, mathematics, architecture, art, and natural systems to address engineering challenges. However, it is crucial to note that engineering neither belongs to a superset nor a subset of any of these disciplines; it possesses distinctive features.

Similarly, philosophy shares common ground with engineering, as it also concerns itself with topics such as basic sciences, mathematics, architecture, art, and natural systems. Nevertheless, philosophy approaches these subjects differently and employs different problem-solving tools. This underscores the unique nature of both disciplines despite their shared areas of interest.

The perspective that "philosophy is not a theory but an activity" [9] demonstrates that the fundamentals of philosophy can be applied to derive a framework of attributes, characteristics, and principles that collectively define engineering's distinct nature and practice. As such, philosophical activities are employed to derive a set of observations that can be explored in engineering, as philosophy, widely accepted as Epistemology, Ethics, Logic, and Aesthetics, can provide valuable insights. Table 1.1 summarises the branches of philosophy in order of relevance to engineering practice, offering a concise description of each branch and examples of categories within each division [9]. The main key steps involved in an engineering practice include the following:

Step 1: Modelling the process or collecting information to acquire relevant knowledge.

Step 2: Formulating the objectives and ideals and applying the acquired knowledge within specific codes of practice in a design process.

Step 3: Recording the outputs or results of implementing the design.

Table 1.1 The main branches of philosophy

Branch of philosophy	Description	Key philosophical approaches	Role in science and engineering
Epistemology	Examines the nature, origin, scope, and process of acquisition of knowledge	Rationalism, empiricism, and logical positivism, falsification	Guides the scientific and engineering methodology and inquiry processes
Ethics	Investigates principles of morality and concepts of integrity and misconduct	Deontology, utilitarianism, virtue ethics with morality, virtue, duty, consequences	Provides ethical frameworks for decision-making in engineering and scientific practices
Logic	Studies the principles of valid reasoning and inference	Deduction, induction, propositions, validity, falsification, and fuzzy logic	Enhances precision in scientific reasoning and problem-solving in engineering
Aesthetics	Examines concepts of beauty, art, and sensory experiences	Formalism (Emphasises form and structure), Realism (Seeks to represent things as they are), Expressionism (Focuses on expressing emotion), Pragmatism (Emphasises practicality and functionality), Postmodernism (Challenges traditional norms)	Influences design aesthetics and user experiences in engineering and technology

Step 4: Based on heuristic rules, personal experience, or mathematically defined performance criteria, the outputs or results are evaluated.

Step 5: The experience or evaluations and verifications are used to enhance the knowledge.

The above closed-loop learning process guides engineers to arrive at a practical engineering solution to a problem. Figure 1.1 depicts this engineering model, with a particular emphasis on activities that are philosophical in nature or can be enhanced by philosophical considerations [9]. The creative design processes crucial to sound engineering practice rely significantly on epistemology, logic, ethics, and aesthetics.

In general, the engineering, science, and technology knowledge base has been shaped by the efforts of engineers, scientists, and technologists. This influence spans the increasingly interconnected global experiences of recent decades and extends back over many centuries. Mathematical or data models, in broad terms, are a significant component of this knowledge, serving as tools to represent and analyse complex systems, make predictions, and guide decision-making processes. They contribute to the broader understanding and advancement of these domains.

Logic is a fundamental principle in engineering, serving a pivotal role in organising complex activities and individual tasks, such as the application of mathematics or the

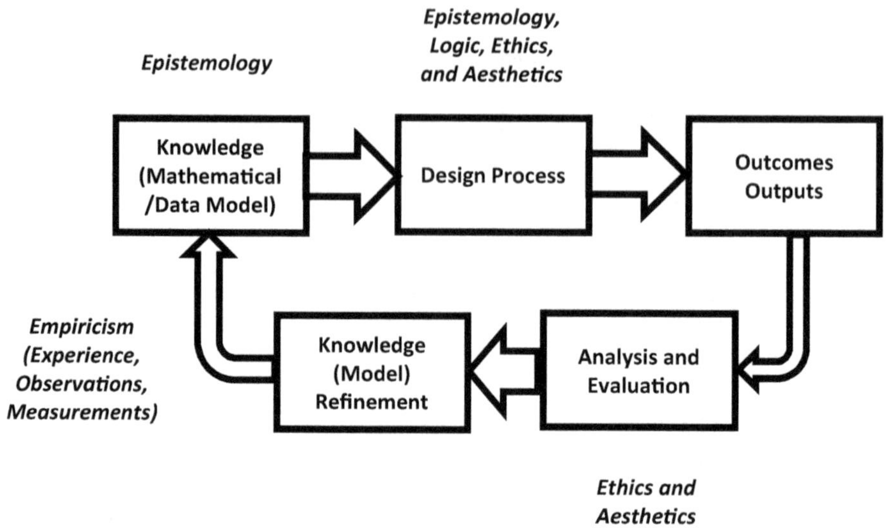

Fig. 1.1 A model of engineering design and its links to philosophy

design of control systems. The significance of ethical practices in engineering and science is increasing, influencing our contemporary living. However, ethical issues will only be briefly addressed later in Chap. 10.

1.2 Feedback and Control Systems Design

This section offers a concise overview of feedback control's role in dynamical systems, complemented by a historical-philosophical discussion of its development and significance.

1.2.1 A General Overview

A *dynamical system* is a system that changes its behaviour with time. The dynamic nature of the system is reflected in the temporal evolution of its internal variables or parameters. This temporal evolution refers to the change or development of a system and describes how the system's internal variables or properties vary as time progresses. The key characteristic is that the system's state at one point in time differs from its state at another. This evolution is governed by rules or equations that describe how the system's internal variables change over time.

Also, the behaviour of a dynamical system is often influenced by external factors in the form of manipulated inputs, disturbances, or environmental changes that impact its dynamics. The behaviour of almost all dynamical systems requires modifications to achieve acceptable performance. Aristotle expresses this nicely and comprehensibly "... *if every instrument could accomplish its own work, obeying or anticipating the will of others... if the shuttle would weave and the pick touch the lyre without a hand to guide them, chief workmen would not need servants, nor masters slaves.*" as quoted from [10].

Control systems design involves systematically applying principles and methodologies to regulate the output or behaviour of dynamical systems in a desired direction. Most control systems designs operate as *closed-loop systems*, while only a few function as *open-loop systems*. At the core of closed-loop control system design are *feedback mechanisms*. Figure 1.2a, b show the general structure of open-loop and closed-loop control systems, where the desired performance is the performance defined by the user for the output to follow.

Feedback mechanisms involve utilising the output of a dynamical system to generate a regulating input signal. Besides, when two or more dynamical systems are interconnected and influence each other with strongly coupled dynamics, simple *causal reasoning* about a feedback system becomes challenging. Simple causal reasoning refers to understanding and explaining cause-and-effect relationships straightforwardly. In the context of feedback mechanisms, the ability to trace and comprehend the direct influence and consequences between different elements or

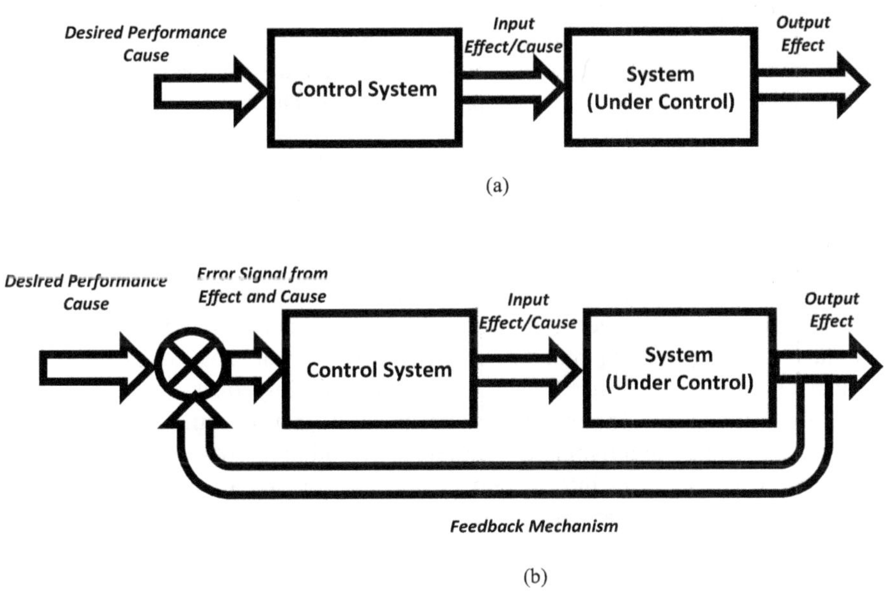

Fig. 1.2 **a** An open-loop control system, **b** A closed-loop control system

components within a system is not straightforward. When a system involves feedback, especially in interconnected and mutually influencing dynamics, the cause-and-effect relationships become more complex, making it difficult to apply simple causal reasoning.

On the other hand, in an open-loop control system, there is a straightforward cause-and-effect relationship between the input and the output, as the output is determined solely by the input without any feedback mechanism. The system's response is predetermined, and there is no adjustment based on the actual output. In this context, the cause (input) directly leads to the effect (output) without being influenced by the system's performance, as shown in Fig. 1.2a. However, in feedback interconnections, the first system influences the second, and the second system influences the first, as shown in Fig. 1.2b. This leads to a *recursive causality*. That is, in feedback mechanisms, understanding the cause-and-effect relationships becomes challenging because the first system's influence is tied to the second and vice versa, creating a loop of interdependence.

The behaviour of feedback systems often deviates from common expectations or intuition, as defined by an error signal shown in Fig. 1.2b. Systematic and structured approaches, such as mathematical or analytical methods, become necessary to gain a deeper understanding of the effects and implications of feedback in closed-loop systems. This is crucial for achieving a more accurate and reliable comprehension of feedback systems [11].

In the modern world, control systems are ubiquitous, essential, and largely invisible. Products, including the stationery used, the water consumed, the food eaten, the cars driven, the drugs and medical equipment that treat diseases, and household utilities, all incorporate simple or complex controllers. The pervasiveness of control systems in daily life highlights control theory's role as a fundamental engineering discipline and an applied science shaping technological progress. As we encounter the complexities of our modern world, the science of control systems forms the fundamental framework for addressing these engineering challenges, enabling current technologies, and driving future developments.

There is an extensive and vast literature on control systems design, where control systems scientists, researchers, and professional engineers explore numerous theories, methodologies, and applications. This extensive body of knowledge explores the fundamental principles governing control systems and investigates advanced techniques for optimising their performance. From classic texts to cutting-edge research papers, the literature highlights the continuous evolution of control systems design as a field. This wealth of information is valuable for practitioners and scholars, providing insights that contribute to the ongoing refinement and innovation within control systems.

1.2.2 *Significance of the Historical and Philosophical Perspectives in Studying Feedback Control Systems*

Naturally, the main body of literature, research, education, and discussions on control systems predominantly concentrates on technical-mathematical aspects, with few exceptions in the form of historical documents and scattered philosophical discussions [12]. However, with the evolution of increasingly complex systems, it is equally crucial to recognise the importance of historical and philosophical discussions on control systems. A holistic understanding of control systems should encompass the technical foundations and the broader historical and philosophical dimensions for a comprehensive grasp of the field.

Studying feedback control systems from a historical and philosophical perspective can be important for several reasons. Examining the historical context in which feedback control systems were developed provides a deeper understanding of the challenges, motivations, and technological constraints that influenced their evolution. This contextual understanding can enhance the comprehension of the fundamental principles and theories. This also reveals the evolution of key concepts, methodologies, and design approaches. Understanding how these ideas have evolved can provide insights into their underlying principles and help practitioners appreciate the real-world challenges that inspired them.

Historical studies often include case studies of successful and unsuccessful applications of control systems. Analysing these cases can provide valuable lessons and insights for current practitioners, helping them avoid pitfalls and make informed decisions in their designs. This could lead to a deeper and more inclusive understanding of feedback control systems from a global perspective, allowing for the investigation of different cultural, economic, and technological influences on their development.

Recognising the conditions and historical circumstances that led to specific advancements in control systems shows the technological progress's non-linear and unpredictable nature. This awareness can influence decision-making in the present research and development. It can also highlight instances of technological convergence, where ideas and technologies from different fields come together. Comprehending these convergences provides a framework for interdisciplinary collaborations and innovations in the present day.

Understanding the historical development of control systems can also involve considering the cultural and societal influences that shaped their applications. Although this aspect of control engineering history has not been deeply studied, the impact of control systems on various industries, economies, and societal structures can be better appreciated through a historical perspective.

Historical studies help identify long-term trends and paradigm shifts in control systems. Recognising these shifts can provide insights into the direction of the field, helping researchers and practitioners anticipate future developments and challenges.

Also, examining the historical development of control systems can provide insights into the human factors and interactions involved. Understanding how user administrators, interfaces, human–machine interactions, and psychological considerations have evolved is crucial for designing user-friendly and effective intelligent control systems. From this perspective, a wartime study of control systems can be found in [13]. Furthermore, historical perspectives on control systems can reveal the policy decisions and regulations that have shaped their deployment. Analysing the consequences of different policy approaches can be valuable to discussing contemporary governance, privacy, and security issues.

The philosophical foundations of control systems, such as the principles of cybernetics and system theory, contribute to the theoretical framework of the field. Exploring these philosophical foundations can help clarify the conceptual basis of control systems and their broader implications. Also, applied philosophical studies can inspire innovation by showcasing how past control scientists and engineers addressed challenges and developed innovative solutions based on certain philosophical concepts. These breakthroughs can stimulate creativity and encourage new approaches to contemporary problems.

Control systems connect to various disciplines, including basic sciences, engineering, and computer science. A historical and philosophical perspective encourages an interdisciplinary approach, fostering a more holistic understanding of control systems.

Also, the study of control systems from a philosophical perspective can raise ethical considerations related to using technology for control and automation. Examining the ethical implications of control systems is fundamental to responsible design practices. This can also be vital with the advent of autonomous and intelligent control systems technology. Questions related to autonomy, human agency, and the ethical dimensions of control can be critically examined through a philosophical analysis.

Finally, incorporating historical and philosophical perspectives into control systems education enhances the learning experience for students. It provides a broader context for theoretical concepts and practical applications, promoting a well-rounded understanding of the subject and leading to a comprehensive understanding that extends beyond technical aspects, encompassing societal, cultural, and ethical dimensions.

1.3 Engineering Ontology: Distinguishing the Interconnected and Divergent Concepts and Methods of Engineering and Science

It is important to distinguish between engineering and science despite the scientific nature of engineering. The fundamental concepts, entities, and relationships within the engineering field, that is, engineering ontology, are different from science. Engineering theories deal with different entities and are judged by different criteria.

A model derived in [14] from Popper's three-worlds ontological framework[3] is proposed that is analogous to the deductive-nomological view of scientific theories, where this view of scientific theories, also known as the hypothetico-deductive model, is a philosophical perspective on how scientific theories are constructed, tested, and used to explain phenomena. This approach was prominent in the logical positivist movement of the early twentieth century. For more details on Popper's three-worlds ontological framework, refer to [15], and for a deductive-nomological view of scientific theories, refer to [16]. The derived modelling framework identifies three key elements from engineering definitions: *requirements*, *designs of artefacts*, and *theories* for reasoning about how artefacts will meet requirements.

Engineering practice in the modern world adapts scientific theories, but not in the same way science or applied sciences do. Engineering epistemology examines the processes of knowledge acquisition, generation, and validation in engineering practice. It discusses how engineers obtain, develop, and apply knowledge to solve problems, design systems, and create innovations. While sharing some similarities with scientific epistemology, it differs fundamentally in its emphasis on synthesis, design constraints, and practical implementation.

According to [14] and aligned with the Engineers' Council for Professional Development (ECPD),[4] engineering can be defined as "The creative application of scientific principles to design or develop structures, machines, apparatus, or manufacturing processes, or works utilizing them singly or in combination; or to construct or operate the same with full cognizance of their design, or to forecast their behavior under specific operating conditions; all as respects an intended function, economics of operation and safety to life and property." The three elements in engineering definitions are artefacts, plants or systems, requirements for these, and theories for predicting whether they will satisfy requirements.

Physical artefacts are the central focus of engineering and can include purpose-built tools and any configurations of matter and energy used for a specific purpose. Examples of artefacts range from structures and devices to system components, systems, and physical processes.

Requirements for artefacts go beyond the primary function of an artefact and encompass acceptable side effects, operational constraints, and environmental conditions. The success of a technological action may depend not only on its main goals but also on the unavoidable side effects and secondary consequences it produces.

Finally, engineering theories are crucial in predicting whether an artefact will satisfy its requirements. Engineering theories include adaptations of scientific theories and empirically validated mathematically represented theories specifically developed within engineering research. However, engineering theories also include rules of thumb, laws of similarity underlying physical models, tables of performance data intended to be suitable for interpolation over a limited range of a few variables, or

[3] Karl Popper's three-worlds ontological framework, also known as the World 1, World 2, and World 3 theory, is a philosophical concept that categorises different realms of reality. This framework helps in understanding the relationships between physical objects, experiences, and abstract concepts.

[4] Engineers' Council for Professional Development was later renamed as ABET.

safety factors that emerge from engineering practice. The characteristics of engineering theories, which have epistemological implications and significant effects on our understanding of knowledge and truth beyond just technical aspects, also distinguish engineering from science and are presented as [14]:

- *Sufficient to be phenomenological*: Scientists aim to create fundamental theories that have ontological and behavioural correspondence with the world, whereas engineers may be satisfied with theories that only predict phenomena without necessarily explaining or having an ontological connection to the world.

For example, consider the design of a temperature control system for a chemical reactor. In this case, scientists and engineers would collaborate to understand the underlying chemical reactions, heat transfer mechanisms, and other factors influencing the temperature within the reactor. The scientists might aim to develop a fundamental understanding of the system by creating a detailed mathematical model that captures the complex interactions between various factors. This model would ideally have ontological correspondence with the real world, explaining the underlying mechanisms and behaviours of the system. On the other hand, engineers working on the temperature control system might focus more on designing a practical and effective control system utilising the PID[5] control methodologies to achieve the desired temperature within the reactor. PID controllers can use simplified models or empirical data tables to predict temperature behaviour under different operating conditions. These models might not have the same explanatory power or ontological correspondence level as the scientific model but would still enable the engineers to design a control system that meets the desired performance criteria. This clearly shows that limited phenomenological theories can provide sufficient information within engineering.

- *Wide variety of levels of precision*: Engineering and scientific theories differ in precision, accuracy, and the role of approximations. Engineering theories can be very precise but often rely on conservative approximations. These approximations are designed to account for various uncertainties and help increase confidence in the accuracy of the claims made by the theory. In engineering, the use of safety factors and idealisations is common to ensure the reliability and performance of an artefact, even in the face of uncertainties. However, these suboptimal design decisions can still be considered good enough according to the *concept of satisficing*, that is, the solutions are adequate for the given situation, even if they might not be the optimal ones. On the other hand, scientific theories aim to push the limits of precision and strive to understand the fundamental truths behind phenomena. In science, theories are tested rigorously to determine if their inaccuracies are due to fundamental flaws, calculation errors, or experimental limitations.
- *Wide variety in limitations on scope*: Engineering theories often have a narrow scope of applicability, being limited to specific systems, classes of designs, or types of behaviours, as they are tailored to address practical problems and provide

[5] Proportional-Integral-Derivative.

solutions within the context of the given constraints. Engineering theories could be more broadly applicable, but their limitations are often justified if they cater to the specific operating conditions acceptable for an artefact's intended use. This approach aligns with the concept of satisficing, where solutions are considered adequate if they meet the required criteria, even if they are not the most general or universal. Various factors, such as the derivation of special cases from general theories or the research methods employed, can influence the scope of engineering theories.

For example, consider the design of a temperature control system for a manufacturing process. The designed PID control focuses on maintaining a specific temperature range for a particular production step. In this case, the engineering theory would involve adjusting the heat input to the system based on the temperature measurements. The theory might be derived from a more general control theory (any of the PID tuning algorithms), considering specific environmental conditions, disturbances, and manufacturing process constraints. These factors influence the controller's design and tuning, making the control structure and tuning highly specialised and tailored to the specific process. The scope of the temperature control theory in this example would be limited to maintaining the desired temperature range within the given manufacturing process. It might not directly apply to other temperature control systems with different requirements or operating conditions.

- *Wide variety of target qualities*: Engineering theories encompass many factors, while science primarily focuses on understanding pure physical phenomena such as space, time, matter, and energy. Modern sciences also explore various other aspects, including psychology and ecology. On the other hand, engineering goes beyond these scientific pursuits by considering additional socio-technical qualities. These qualities are not part of traditional scientific inquiry and include factors like cost, time, safety, and usability of an artefact. Engineering theories must address these socio-technical aspects, as they often play a crucial role in the design, construction, and operation of artefacts.
- *Wide variety of instrumental properties*: The instrumental properties of engineering theories are of significant importance in the field, as they must be practical, efficient, and applicable within the limitations of available resources. This contrasts with the scientific community's focus on simplifying complex problems for analysis, where the difficulty of understanding and using theories may not hold the same significance level.
- *Multiple overlapping theories*: Scientists typically focus on a single theory or a few competing alternatives with the broadest scope and highest precision. In contrast, engineering theories often encompass multiple, unfalsified theories with varying precision, applicability, and instrumental properties. Engineering knowledge at the forefront of a domain may consist of overlapping theories, each excelling in one or a combination of these characteristics. This allows for trade-offs in selecting theories, as an engineering theory is not necessarily rejected unless it fails to correspond with reality or is outperformed by another theory across the range of these characteristics.

An example of the multiple overlapping theories is the use of engineering rules of thumb, which may have a narrow scope of application and coarse level of precision compared to other alternatives. However, they remain part of the engineering body of knowledge due to their cost-effectiveness and ease of use.

In applying the concepts and methodologies of the philosophy of science to engineering, the differences between engineering and science must be accounted for. Engineering, in general, encompasses the fields of its practice and implementation, research and methodology, or the engineering profession. However, engineering research and methodology are the closest to science, so unless otherwise specified, engineering is considered engineering research and methodology.

Historically, technology (including engineering) was considered synonymous with applied science. However, recent perspectives from historians and philosophers have identified some key differences between the two. Both applied science and engineering share common elements, such as drawing on general scientific theories and developing specialised, narrow-scope theories that would not be acceptable in pure science. They also both focus on understanding phenomena within the context of specific requirements. Despite these similarities, there are distinct differences. While both applied scientists and engineers, analyse designs for artefacts, engineers tend to be more concerned with creating and realising those designs. They often have legal privileges and responsibilities in assuring artefacts' safety and reliability. Additionally, engineers typically exhibit greater concern for the discovery and characterisation of requirements than applied scientists.

Engineering does not always apply science, and engineering discoveries may precede scientific explanations. There are several examples in control systems engineering where engineering discoveries preceded scientific or mathematical explanations or where engineering principles have significantly impacted scientific understanding. Proportional control is one of the earliest control system concepts, where the control signal is directly proportional to the error between the desired output and the actual output. This concept was widely used in engineering applications before being formally explained by the scientific theory of feedback control systems.

The Ziegler-Nichols tuning method is another example of industry developments that precede scientific explanations. Although the Ziegler-Nichols method was developed in industry, it later gained scientific grounding by studying its underlying principles and developing a more advanced control theory. The method remains a valuable tool for tuning PID controllers in practice, especially when a more sophisticated approach might not be necessary or feasible.

Model Predictive Control (MPC) is another example of advanced control. It is a strategy that uses a mathematical process model to predict future behaviour and make optimal control decisions. This engineering technique was developed and applied in practice before being thoroughly studied and understood from a scientific perspective in control systems.

Finally, the Relative Gain Array (RGA) method for input–output pairing can also be considered an example of multivariable control systems design. The RGA method was initially developed in the process industries, particularly in chemical plants, as a practical approach to finding an input–output pairing for multiple feedback loops in a process, ensuring that the overall system stability and performance are improved.

Engineering knowledge creation in both research and design contexts is closely interrelated. Designs can vary in generality, from highly specific to highly parametric, but these designs may lead to new design principles applicable to a broader range of problems. Although engineering practice primarily focuses on designing and analysing specific artefacts, knowledge creation still occurs within this context, particularly concerning new designs and their relationships with specific artefacts and requirements.

1.4 A General Systems Classification

The systems encountered in control engineering can vary in their level of uncertainty (aleatory and epistemic), interconnectedness and interactions, degree of nonlinearity, emergence behaviour, and control difficulty. In a general system classification, four system types that encompass the systems encountered in control systems design could be identified. This classification includes [5] *simple, complicated, complex, and complex adaptive systems.* The proposed classification is shown in Fig. 1.3, which indicates that the boundary between these system types is not very well defined and is fuzzy. The following characterisations are presented to explicitly define each of these systems. This classification is used throughout the book to distinguish between various encountered systems.

Simple systems have one input and one output with internal dynamics that can be modelled by either linear differential equations or linearised with good accuracy governing system behaviour. Examples of simple systems can be a DC motor with input voltage and rotational speed as its output or the level control of a tank filled with water with a valve to change the input water inflow.

Systems with multi inputs and multi outputs consisting of several simple subsystems can be simple systems if they exhibit weak or no interconnectedness or interactions. *Complicated systems* are systems with multi-inputs and multi-outputs and strong interactions between subsystems and a possible uncertainty that can be modelled with bounds. However, these interactions do not lead to emergence. Examples of complicated systems can be a distillation column in the process industry or the control of the angles of an aircraft with ailerons.

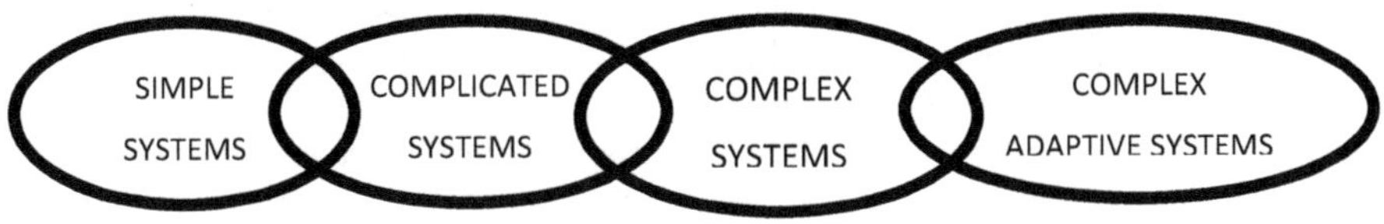

Fig. 1.3 A general systems classification

Emergence occurs when a system property is manifested, and its subsystems do not have that property. However, some complicated systems can exhibit *domino behaviour*, where a failure in any of the subsystems leads to total or partial system failure. The degree of failure would depend on the interaction measure. Although the domino behaviour occurrence has a low probability, it can cause irreversible and serious consequences with strong negative effects.

Complex systems are (uncertain) systems with strong interaction among their subsystems, where the interactions or interconnectedness can lead to emergence. Note that single input and single output complex systems exist with emerging system behaviours. Examples of complex systems can be ant colonies where individual ants follow simple rules, but collectively they can exhibit complex behaviours, neural networks with collective behaviour of interconnected neurons in the brain resulting in complex cognitive processes such as learning, memory, and decision-making. Traffic flow is another example where the interactions between individual vehicles on a road can lead to emergent phenomena such as traffic jams and congestion despite each vehicle following simple driving rules.

Finally, *complex adaptive systems* are systems with strong interactions among their subsystems, and their subsystems(agents) can have time-varying characteristics, or they can adapt their behaviour to changes in the system, be it changes in the environment or the behaviour of other agents. Human society, financial markets, social networks, and smart power grids are all examples of a complex adaptive system.

1.5 A Note on Complexity and Complex Adaptive Systems

Murray Gell-Mann, a Nobel laureate in physics, discussed the etymology of the concept of complexity in [17]. The Latin root *plexus* means *braided* or *entwined*, and it is related to the concept of complexity. The Latin word *complexus* is derived from *plexus*, which means *braided together* or *entwined*. This Latin word has been borrowed into various languages, including English, where it has evolved into the word *complex*. In English, the word complex initially referred to something intricately entwined or braided, such as a complex knot or pattern. Over time, the word's meaning has expanded to include a broader sense of interconnectedness or interdependence between different parts or components. The importance of interactions and feedback loops within systems is emphasised in the modern scientific understanding of complexity. Gell-Mann indicates that it is common to define the individual parts or aspects in advance and then examine them carefully when studying complex systems. However, even when experts combine their results, an adequate description of the whole system or situation often does not emerge. This is because these parts or aspects are typically interconnected, meaning they influence and affect each other in complex ways. Complexity practitioners often adopt a transdisciplinary perspective to overcome this limitation, taking a holistic approach considering the entire system.

This helps to reveal the intricate relationships and patterns that underlie the complex behaviour.

Complex adaptive systems (CAS) consist of diverse *components* that interact and influence each other in complex ways. These components, often referred to as *agents*, can exhibit emergent behaviour at the system level, and they can learn from experience and adapt to change in their environment. Engineering examples of CAS include Smart Grids, Autonomous Vehicle Networks, Industrial IoT Systems, Cyber-Physical Systems, Swarm Robotics, Biome-inspired Systems, and Electric Vehicle Charging Networks, Refer to Chap. 10 for specific examples and the control challenges.

Complexity Theory and Complex Adaptive Systems have significantly advanced the understanding of physical and social systems, challenging traditional linear and reductionist approaches. While computer-based modelling, such as agent-based simulations, has been a key factor in this development, other methods have also contributed to our understanding.

Despite the significant advancements in understanding CAS through Complexity Theory and computational modelling, traditional control approaches may not be sufficient to effectively control and optimise these systems. The complexity and uncertainty inherent in CAS, coupled with the emergent behaviour, can lead to suboptimal performance, instability, and even catastrophic failures. Moreover, the rapidly changing environment and dynamic interactions between agents can render traditional model-based control approaches inadequate, as they are often based on simplified assumptions and linearised models.

In contrast, data-driven control approaches can provide a more robust and adaptive solution to control CAS. By leveraging data collected from the system and its environment, data-driven control approaches can learn and adapt to changing system dynamics, optimise system performance, and provide real-time feedback to improve system stability and resilience. For more details on the role of data-driven control methodologies in CAS, refer to Chaps. 8 and 10.

1.6 Outline of the Book

The book is organised into ten chapters. This chapter is the Introduction. This Chapter provides a perspective on the importance of history and philosophy to engineering. It then emphasises the role of history in engineering practice and philosophical studies in engineering design. The Chapter introduces control systems and the historical-philosophical studies of these systems design. Control systems analysis and design are the main subjects of this book, and all the other materials presented are related to them.

Chapters 2–4 provide a comprehensive history of control systems design. It is to be noted that [10, 18, 19], are the presently available books on the history of control systems. However, each book deals in detail with a specific period in control systems. A more comprehensive history of control systems up to modern times can be found in [20] and the introduction of several textbooks. See, for example, [12, 21].

Chapter 2 deals with control systems from antiquity to the pre-classical control era. The earliest developments in automatic control, which often consisted of simple mechanisms and devices created by ancient civilisations are briefly discussed. The developed versions of these systems through to the pre-industrial era and leading up to the Industrial Revolution are then presented. The pre-classical control era and the period before the formal establishment of systematic control principles, including advancements in water clocks, automata, and other early control-related technologies, are addressed.

Chapter 3 explores the classical to modern control period. The key point emphasised is the formalisation of control principles and theories, the systematic study of feedback control systems, and the development of fundamental control theory. On the firm-established feedback theory of the early twentieth century arose modern control techniques rooted in the developed mathematical systems theory.

Chapter 4 briefly presents the historical developments in the postmodern control approaches for handling uncertain and unknown complex systems. These approaches include robust, adaptive, and intelligent control techniques.

Models and modelling are fundamental in modern and postmodern control design methodologies. Chapter 5 presents the philosophical foundation of mathematical modelling and explores its role and limitations for applying advanced mathematical tools in control system analysis and design. Key issues such as the certainty equivalence principle, model uncertainty, and the model-assumption dilemma are also addressed. Uncertainty as a bottleneck of the model-based control systems design is further discussed in terms of the trio of uncertainty combat: *Robustness*, *Adaptation*, and *Intelligence*.

Chapter 6 deals with system theory. This groundbreaking multidisciplinary theoretical framework provided an effective tool to describe and analyse complex systems, regardless of their specific nature or discipline. The philosophical principles and background, as well as their implications for control engineering, are discussed in this Chapter. Then, mathematical system theory is presented and discussed as a major advancement in control theory. The limitations and some critical views on mathematical theory are provided at the end of Chap. 6.

Model-based control design techniques are primarily induction-based approaches. In Chap. 7, it is argued that the principles of falsification philosophy have influenced various aspects of control systems design, including system identification and modelling, controller design, performance testing, fault tolerance, and safe control. This is a notable example of the direct implication of a philosophical concept in control systems design, which resulted in the unfalsified adaptive control and provided a solid framework for data-driven control methodologies.

Chapter 8 discusses paradigm shifts in control systems through Kuhn's philosophy of science frameworks. It is shown that based on Kuhn's theories on scientific revolutions and paradigm shifts, three paradigm shifts are observed, and a fourth paradigm shift is on the verge of occurring. This Chapter employs historical pieces of evidence and technical arguments to specify these paradigm shifts within control systems design. Kuhn's theories of scientific revolutions and paradigm shifts are utilised to develop a model for understanding these shifts in control systems design.

The effect of modernism on engineering, particularly control systems design, is studied in Chap. 9. It is discussed that modernism has shaped the evolution of control methodologies and theoretical frameworks with mathematical formalism and optimality. Also, postmodernism's effect on control is discussed and observed as a response to what is perceived as limitations of modernist thought and embracing a more pluralistic and relativistic view.

Finally, Chap. 10 reviews the challenges, advancements, and applications of control theory and practice over time, envisioning the coming years as perceived by prominent control scientists and engineers from an early 1963 paper up to the road map published in 2023. These scientists and engineers have reflected on the evolution of control systems from the past to the present and into the future within their respective eras.

References

1. https://www.abet.org/accreditation/accreditation-criteria/criteria-for-accrediting-engineering-programs-2022-2023/. Accessed 14 Jan 2025
2. https://theconversation.com/why-science-and-engineering-need-to-remind-students-of-forgotten-lessons-from-history-61356. Accessed 14 Jan 2025
3. A. Gaynor, G. Crebbin, What can engineers learn from the past? A potential role for history in engineering education. Int. J. Eng. Soc. Justice, Peace **2**(2), 43–54 (2013)
4. A. Picon, Engineers and engineering history: problems and perspectives. Hist. Technol. **20**(4), 421–436 (2004)
5. D. Michelfelder, N. Doorn, *The Routledge Handbook of the Philosophy of Engineering* (Routledge-Taylor & Francis Group, 2021)
6. J. Hillier, The philosophy of engineering: a critical summary, in *Proceedings of the Institution of Civil Engineers-Civil Engineering*, vol. 162, no. 2 (Thomas Telford Ltd, 2010), pp. 91–95
7. D. And one, D. Merezeanu, Modern power plant control philosophies, in *CSCS–12, 12th International Conference on Control Systems and Computer Science*, vol. 1 (1999), p. 6
8. C. Mitcham, The importance of philosophy to engineering, Teorema: Revista internacional de filosofía 27–47 (1998)
9. W. Grimson, The Philosophical nature of engineering—a characterisation of engineering using the language and activities of philosophy, in *2007 Annual Conference & Exposition* (2007), pp. 12.1453.1–12.1453.14
10. S. Bennett, *A History of Control Engineering, 1930–1955* (no. 47) (IET, 1993)
11. K.J. Åström, R.M. Murray, *Feedback Systems: An Introduction for Scientists and Engineers* (Princeton University Press, 2021)
12. A. Khaki-Sedigh, *An Introduction to Data-Driven Control Systems* (Wiley-IEEE Press, 2023)
13. D.A. Mindell, Engineers, psychologists, and administrators: control systems research in wartime, 1940–45. IEEE Control Syst. Mag. **15**(4), 91–99 (1995)
14. M. Staples, Critical rationalism and engineering: ontology. Synthese **191**(10), 2255–2279 (2014)
15. B. Boyd, Popper's world 3: origins, progress, and import. Philos. Soc. Sci. **46**(3), 221–241 (2016)
16. J. Woodward, E.N. Zalta, *Scientific Explanation* (The Stanford, 2017)
17. M. Gell-Mann, Complexity at large. Complexity **1**(5), 3–5 (1996)
18. O. Mayr, The origins of feedback control. Sci. Am. **223**(4), 110–119 (1970)
19. S. Bennett, *A History of Control Engineering 1800–1930* (The Institution of Engineering and Technology, 1979)

20. A. Khaki-Sedigh, *History of Control Engineering* (K. N. Toosi University of Technology Press, (In Persian), 2014)
21. F.L. Lewis, D. Vrabie, V.L. Syrmos, *Optimal Control* (John Wiley & Sons, 2012)

Chapter 2
A Concise History of Control Systems: From Antiquity to the Pre-classical Control Era

> *The history of science shows that theories are perishable. With every new truth that is revealed, we get a better understanding of nature and our conceptions and views are modified.*
> *—Nikola Tesla*

2.1 Introduction

The study of the history of control systems design is important for several reasons, as discussed in Sect. 1.2.2. It reveals the progression of control scientists' and engineers' knowledge and understanding of the *feedback* in man-made and physical systems and provides insights into how control theories and design concepts have developed over time. The history of control theories and design concepts is intertwined with philosophical changes. It reflects how broader philosophical viewpoints have influenced scientific thinking. It can inspire current and future control systems theorists and engineers to tackle challenging problems and observe them from a broader outlook.

Control theories and design concepts have experienced paradigm shifts, where fundamental theories and practices have been replaced by new ones that drastically altered the system analysis and design perceptions. Studying these paradigm shifts can provide insights into how control systems science evolves and how new ideas gain acceptance.

Finally, control systems education can benefit from understanding how new concepts have evolved and communicating complex ideas more effectively to beginners in the field.

The term *feedback* was first used in engineering and systems theory. Its origins can be traced back to the work of Harold Stephen Black, who introduced the concept of *negative feedback* in the 1920s and 1930s.

Black published a paper titled *Stabilised Feedback Amplifiers* in The Bell System Technical Journal, in which he introduced the concept of using feedback to stabilise and improve the performance of electronic amplifiers [1]. Black's work laid the foundation for the understanding and application of feedback in various fields, including electronics, control systems, and later in areas such as biology, psychology, and

A. Khaki-Sedigh, *Control Systems: A Historical and Philosophical Perspective*,
https://doi.org/10.1007/978-981-96-7426-8_2

economics. The concept of feedback has since become a fundamental principle in understanding and engineering complex systems and has been applied across various disciplines to describe the cyclical flow of information that influences a system's behaviour.

For example, in biological systems, feedback is evident in processes like homeostasis, where a stable internal environment through feedback is achieved in living mechanisms. In economics, the idea of feedback can be seen in how changes in supply and demand affect prices and, in turn, influence future supply and demand. The idea of feedback in communication and language can refer to how speakers adjust their language based on the listener's response. This reciprocal communication is a form of feedback. Even in creative fields like art and craftsmanship, the idea of feedback is apparent. Artists and artisans often adjust their work based on their own observations or the reactions of others, leading to iterative improvements.

Philosophers throughout history have explored ideas related to feedback in various contexts, including the concept of *circular causality* and the interplay between *cause* and *effect*.

Although the term feedback was not commonly used before the 1920s and 1930s in the specific technical sense it is known today, the underlying principles and ideas have been present in different forms across various disciplines. The development of the engineering and systems theory concept during the early twentieth century marked a formalisation and systematic exploration of feedback's potential in various applications.

This chapter and the following two chapters provide a brief introduction to the history of control systems analysis and design methodologies. The historical timeline of automatic control is divided into the following main periods:

- ***Antiquity to Pre-industrial Period***: This period encompasses the earliest developments in automatic control, which often consisted of simple mechanisms and devices created by ancient civilizations. It spans from antiquity to the pre-industrial era, leading up to the Industrial Revolution.
- ***Pre-classical Control Era***: This period comprises the time before the formal establishment of systematic control principles, which began during the Industrial Revolution.
- ***Classical Control Era***: This era involves formalising control principles and theories. It includes the study of feedback control systems and the development of fundamental control theory.
- ***Modern Control Era:*** This era reflects the advancements made in the twentieth century, where mathematical system theory was developed, and rigorous mathematics was employed to analyse and design control systems. It includes the use of computers and digital technology to improve control systems.
- ***Postmodern Control Era***: This period represents recent and ongoing developments in control systems design. It deals with the control of complex systems utilising optimisation, robustness, adaptation, and intelligence concepts.

This chapter highlights major events in the development of control systems analysis and design methodologies by referencing the historical timeline up to the classical control era.

2.2 Antiquity to Pre-industrial Period

This period refers to the earliest advancements in automatic control and is divided into two distinct phases. The first phase spans from approximately 270 BC to 800 AD, while the second phase covers the period from 800 to 1200 AD.

2.2.1 The Earlier Attempts

It is generally believed that the initial attempts to design control systems trace their origins back to approximately 270 BC, when humans embarked on the ambitious task of standardising time measurement [2]. This attempt, driven by the need to regulate various aspects of daily life and societal activities, laid the foundation for more sophisticated control concepts. As a concrete outcome of these efforts, the Greeks and Egyptians ingeniously constructed water clocks, devices that connected the water flow to measure time intervals. The ingenious design principles formulated during this era paved the way for a remarkable technological heritage that endured for centuries. These principles were not confined to horology but found application across many different disciplines, shaping the evolution of control systems and influencing the advancement of automation in diverse fields. Thus, the insights gained from these ancient endeavours continued to influence the field of control well into the future.

Water clocks operated by measuring water flow from one container to another at a relatively constant rate. They were an important advancement in early timekeeping and had practical applications in daily life, including regulating the length of speeches, determining legal time limits, and indicating hours during the night. A sample water clock is shown in Fig. 2.1.

The key principle of a water clock was the controlled flow of water from one container (the reservoir) to another (the receiving vessel). The flow rate was regulated by the size of the orifice or tube through which water flowed. The smaller the orifice, the slower the flow, and vice versa. Many water clocks had graduated markings on a receiving vessel, allowing users to read the passage of time by observing the water level's rise.

Feedback mechanisms were used in more advanced water clocks to regulate water flow and improve timekeeping accuracy. One common type of feedback mechanism involved maintaining a balance between the inflow of water into the clock and the outflow from it. As water flowed into the clock's reservoir, it raised the water level,

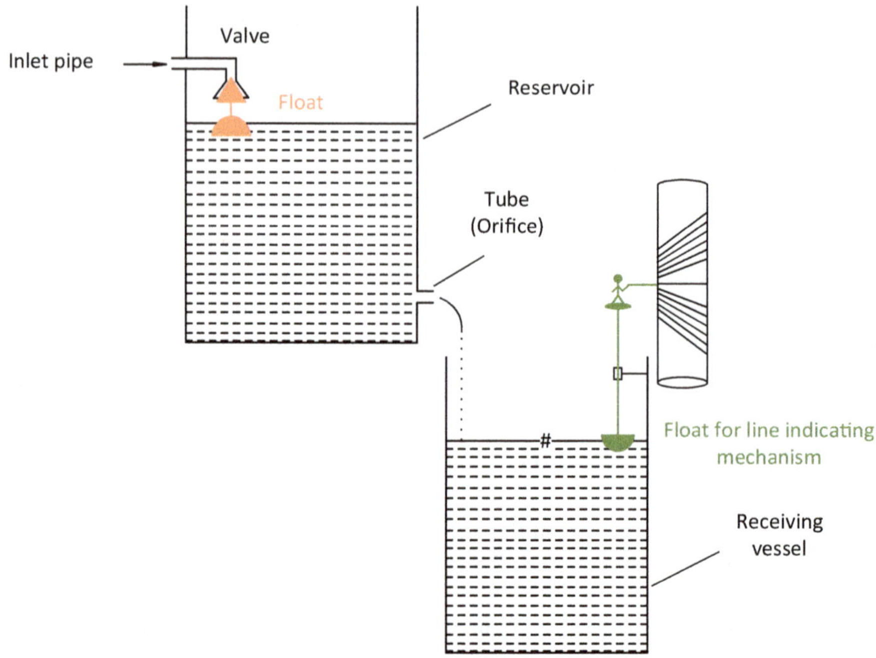

Fig. 2.1 Control of water level by a float

increasing the pressure and controlling the flow rate through an outflow mechanism. This could be a controlled valve or a float-operated mechanism.

Also, a float, often a hollow container or an object with buoyancy, would rise as water filled the reservoir. As the float reached a certain level, it engaged a valve mechanism controlling water outflow. The valve would adjust based on the float's position, ensuring the water flow remained relatively constant and maintaining a consistent water level in the reservoir. These feedback mechanisms aimed to compensate for variations in water pressure, temperature, and other factors that could affect the flow rate. By creating self-regulating systems, water clocks equipped with feedback mechanisms could provide a more accurate measure of time than simpler models that relied solely on gravity and water flow.

A notable figure in the design of water clocks is the ancient Greek Artificer, Ctesibius. He introduced mechanisms such as water *escapements* and *feedback systems* to improve accuracy. An escapement is a mechanism that controls the release of energy in a regulated and intermittent manner as shown in Fig. 2.1.

Ctesibius introduced the concept of a constant-level reservoir in water clocks, using a float-regulated valve system to maintain steady water flow. This reservoir ensured that the water level within the clock remained relatively stable, which was crucial for maintaining a consistent flow rate. The constant-level reservoir operated based on the principle of feedback. For more details and a block diagram of the Ctesibius water clock in Fig. 2.1, refer to [3].

The earliest known written document related to control and servomechanisms is attributed to Heron of Alexandria. As an artificer, Heron's role would encompass both the craftsmanship of constructing devices and the ingenuity of inventing new mechanical solutions. Heron is now believed to have lived in the first century AD, based on a discovery in 1938 that revealed his reference to a *recent* eclipse. This eclipse was identified as the one that occurred in Alexandria at 23:00 h on March 13, 62 AD [4].

It can be inferred from Heron's writings that he was an instructor at the Museum in Alexandria. He authored numerous publications, some as lecture notes for his courses and others as textbooks. Within two of his books, Heron extensively explored the concept of servomechanism systems. One of these works is *The Pneumatica*, consisting of two books that examine mechanical devices powered by air, steam, or water pressure. The second is *The Automaton Theatre*, a puppet theatre operated using strings, drums, and weights.

In his work, *Pneumatica*, presented across two books with 43 chapters in the first and 37 chapters in the second, Heron introduces theoretical and practical fluid pressure considerations. He then presents a comprehensive collection of mechanically ingenious toy-like mechanical systems. This compilation encompasses descriptions of more than 100 machines, including a fire engine, a wind organ, a coin-operated machine, and an *aeolipile*—a steam-powered engine. Notably, Heron's aeolipile, which has much in common with a jet engine, is regarded as the earliest documented device capable of converting steam into rotary motion [4].

Automatic doors originated in ancient Greece and were invented by Heron of Alexandria using a series of ropes and pulleys is one of the earliest examples of a designed control system. This device was used to open the doors of a notable or sacred place and is also known as *Greek automatic doors*. The process begins with a command input, the act of lighting a fire upon the altar. As the fire heats the air underneath it expands, forcing water out of a container and into a bucket. As the weight of the bucket increases, it descends, causing the door spindles to turn through a system of ropes. This action simultaneously raises a counterweight. Conversely, extinguishing the fire initiates the door-closing sequence. As the air within the container cools and pressure decreases, water is drawn back from the bucket into the storage container. Consequently, the bucket becomes lighter, causing the heavier counterweight to descend and effectively closing the door. This cycle continues as long as the bucket remains positioned higher than the container. The device was likely activated as the ruler and his associates commenced their ascent of the temple steps. Notably, the intricacies of the door-opening mechanism were concealed from public view and understanding. This deliberate obscurity cultivated an atmosphere of mystery and highlighted the perceived power of the Olympian gods [5, 6].

Another example is the *south-pointing chariot*, also known as a *south-pointing cart* or *south-pointing carriage*, a navigational system. The south-pointing chariot can be considered an early ancestor of modern navigational systems. This ancient Chinese mechanical device is designed to indicate the south direction regardless of the chariot's actual orientation. The south-pointing chariot was primarily used for navigation and geographical orientation, particularly during travels and military campaigns where maintaining a sense of direction was crucial. This device could help travellers or armies avoid getting lost, especially in unfamiliar or featureless terrains.

The basic principle behind the south-pointing chariot was using gears and a mechanical linkage. The chariot's wheels were connected to a central gear mechanism, which, in turn, was connected to a figurine or pointer on top of the chariot. This pointer would always face south, regardless of the direction in which the chariot turned.

The south-pointing chariot included a feedback mechanism in its design. The feedback mechanism in the south-pointing chariot involved gears and a clever mechanical linkage. The chariot's wheels were connected to a central gear mechanism, and this central gear was also connected to the figurine or pointer on top of the chariot. As the chariot turned and its wheels rotated, the central gear system would adjust the orientation of the pointer to counteract the chariot's changes in direction. This adjustment ensured that the pointer continued to point southward. This feedback mechanism can be seen as a form of closed-loop control [7].

2.2.2 *From 800 to 1200 AD*

During this period, science and technology flourished in Muslim countries. This section briefly reviews the contributions of this era to control system designs. Control systems were instrumental in the development of water clocks. The *on–off control technology* was developed, documented, and employed centuries later in more advanced applications [8, 9]. The control systems technology that emerged in the Islamic world was built upon previously developed techniques. Translated books into Arabic significantly aided Muslim artisans and craftsmen in designing and further advancing control system technologies. One notable example is the translation of Heron's *Pneumatica* into Arabic, which served as a key reference book in the principles of automatic control.

Clocks and timekeeping represented some of the most remarkable achievements of Islamic civilization. During this era, scholars, artificers, and craftsmen engaged in careful mathematical analyses, crafted intricate clocks, and endeavoured to master the principles of automatic control. These efforts led to the creation and refinement of various timekeeping inventions For details and several key examples, refer to [10].

The works and innovations of the Islamic era were documented in several books, clearly demonstrating the principles and application of automatic control in various artefacts. For more details on these books, refer to [9, 11, 12] and the references therein.

A notable example is *Kitab Al-Hiyal* (The Book of Ingenious Devices), a book on mechanical systems written by Banu Musa bin Shakir in the ninth century. This book is based on the original work at the *Bayt Al-Hikma* (House of Wisdom), the leading scientific academy in Baghdad and one of the greatest scientific institutions since the Museum and Library of Alexandria. The Banu Musa were three brothers who supported the translation movement, which gained momentum during the Islamic scientific awakening in the ninth century. They authored over 20 known works, including the seminal engineering book *Kitab Al-Hiyal*.

Banu Musa described one hundred ingenious devices in *Kitab Al-Hiyal*. While some of the devices presented in the book resembled those found in previously translated works into Arabic, there were also Banu Musa machines that bore no resemblance to any of the devices in the translated references.

Banu Musa primarily employed the principles of hydrostatics and aerostatics in their work. They utilised automatic valves and delayed-action systems, and their application of automatic control principles reflects a highly creative approach. Notably, these devices marked the first historical use of crankshafts in technology. Among the numerous pioneering innovations by Banu Musa, one of the most significant is utilising *on–off control* with upper and lower limits for the controlled variable. Systems of this class are widely employed in modern control technology. Originally used by Banu Musa and other Muslim artisans, the float valve experienced a resurgence in Europe and England in the mid-eighteenth century [9].

In [13], the devices of the *Al-Hiyal* book are analysed, and it is shown that the mechanisms employed are examples of actual control engineering practice in the ninth century. For more detailed interpretations of *Al-Hiyal* refer to [14].

The other influential book written by Al-Jazari, another ingenious artificer, in 1206 is *Al-Jami' bain al-'Ilm wal-Amal al-Nafi' fi Sina'at al-Hiyal* (A Compendium on the Theory and Practice of the Mechanical Arts). The distinctive feature of this book lies in its practical nature, enriched with detailed descriptions of various kinds of devices. It is claimed by Hill in [13] that: "It is impossible to overemphasise the significance of Al-Jazari's work in the history of engineering. Until modern times, no other document from any cultural area has provided a comparable wealth of instructions for designing, manufacturing, and assembling machines. Al-Jazari not only assimilated the techniques of his non-Arab and Arab predecessors but also displayed creativity by adding numerous mechanical and hydraulic devices. The impact of these inventions can be traced in later developments such as the design of steam engines and internal combustion engines, thus paving the way for automatic control and other modern machinery. The influence of Al-Jazari's inventions continues to resonate in contemporary mechanical engineering."

A notable example is introducing the *tracking concept* into previously designed water clock control systems [15]. The control systems invented by Ctesibius were regulatory systems, as described in Sect. 2.2.1. Al-Jazari employed the same principles in his water clock systems. However, he was further interested in dividing the time between sunrise and sunset into twelve equal parts following the Islamic principle for prayer times. As this time varied with seasons, Al-Jazari introduced a *reference regulating mechanism* to the water clock control system. This can be considered as the first *set-point-tracking system*. A schematic of this system is shown in Fig. 2.2. For a detailed technical account of the originally designed water clocks refer to [16].

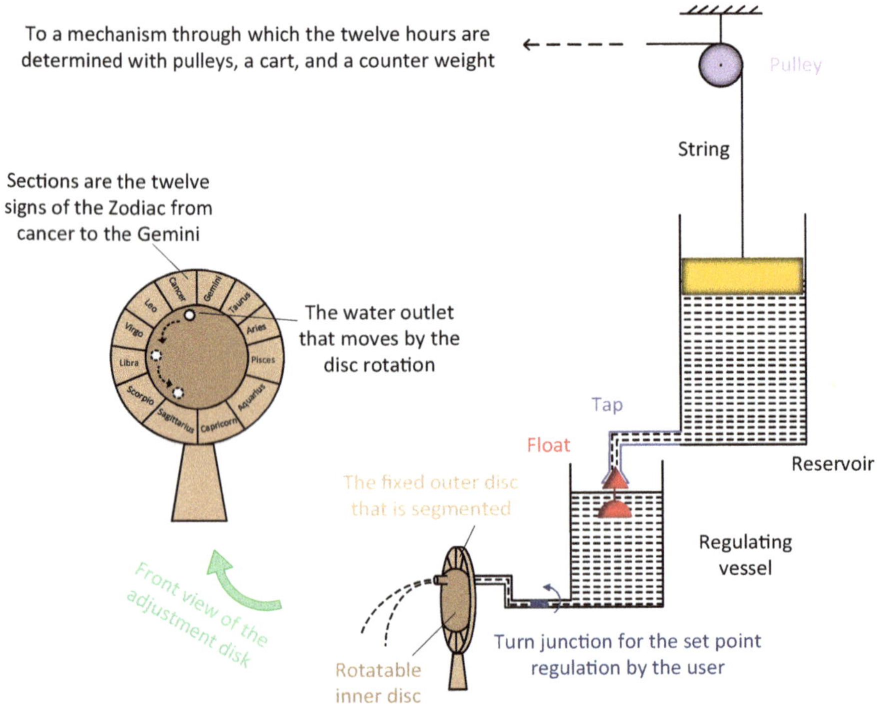

Fig. 2.2 Al-Jazari's water clock and set-point regulating flow rate control mechanism

However, as noted by [3], Ctesibius *might* have used a more primitive method to divide the hours as one-twelfth of the interval between sunrise and sunset. This could have been achieved by varying the rate of change of the water level in the reservoir by adding or removing a wedge inserted in the orifice, as shown in Fig. 2.1, which caused difficulties [3].

A reservoir stores the necessary amount of water for operating the clock mechanism over a day or twelve hours. When the water flows out at a constant rate, the water level in this vessel indicates tracking time from sunrise or the time remaining until sunset.

The water level in the reservoir connects to the clock screen via a rope attached to a float. A counterweight on the opposite end of the rope ensures the necessary friction for rotating the clock screen. At sunrise, the valve at the base of the reservoir is opened, initiating the flow into a regulating vessel equipped with a float valve.

Since the outlet pipe of the regulating vessel is narrower than the inlet pipe, the water level rises, causing the float valve to block the inlet. This leads to a rapid establishment of a steady state condition where the inflow and outflow rates become equal. However, the flow of water from the reservoir and the water level in the regulator depend on the difference between the float valve's position and the outflow pipe's location.

To accommodate this, the outflow pipe can freely rotate around the centre of an adjustment disk, indicating the positions required based on the day or season. During the longer summer days, the exit pipe points upward, while it points downward on the disk during shorter winter days. Consequently, the water level in the regulator tank adjusts with the seasons, influencing the water's outflow time.This represents a change in the reference value, a fundamental concept in control systems.

2.3 Pre-classical Control Era: Leading to the Twentieth Century

The pre-classical control era from approximately 1700s to the early twentieth century marked a significant evolution in control systems engineering, particularly with the onset and progression of the Industrial Revolution.

The 1700s witnessed the development of various mechanical control systems, often driven by the need to regulate machinery in industrial settings. During this period, engineers and inventors began to grasp the concept of *feedback control*, even though it was not yet formally defined. Precision clock making was another area where control systems engineering played a role. The development of accurate timekeeping mechanisms involved intricate control systems involving escapements and balance wheels. In various industrial processes, control systems were applied to regulate temperature, pressure, and flow.

Control systems of this era often relied heavily on *skilled craftsmanship*. Precision engineering and meticulous *workmanship* were essential for creating reliable and accurate control mechanisms.

While the formal field of control theory and engineering had not yet fully emerged, the concepts and practical knowledge developed during this period laid the foundation for modern control systems engineering. In this section, a brief account of selected achievements is presented. This period is of particular importance as it almost coincides with the first and second industrial revolutions, which were a transformative period in human history marked by the shift from manual labour to machine-based manufacturing and the development of new technologies and industries.

It is generally believed that the first industrial revolution occurred between 1760–1840 and began in Britain and spread to other parts of the world, including Europe and North America. It is characterised by mechanisation and the introduction of machines which replaced human labour and increased efficiency, the textile industry, the discovery of new sources of coal and iron, transportation, and the establishment of factories, which replaced traditional cottage industries and allowed for mass production.

The Second Industrial Revolution began in the late nineteenth century and continued until the outbreak of World War I. Electrification, steel production, the automobile industry, the chemical industry, and globalisation characterise it.

These two industrial revolutions profoundly impacted human history, and control engineering played a key role in their achievements.

2.3.1 The Windmills

The use of wind power dates back thousands of years, and evidence of this can be found in various parts of the world, including Iran and China. These early wind-driven devices were simple and often featured vertical-axis designs representing an early form of harnessing wind energy. Vertical-axis windmills, sometimes called Iranian windmills, originated in Persia (modern-day Iran) around 3000 years ago. These early windmills were primarily used for grinding grain and pumping water.

The technology of windmills spread to other regions and eventually made their way to Europe. Windmill designs were adapted and modified, with some transitioning to horizontal-axis windmills with a slight slope. These horizontal-axis windmills became more prevalent in Europe [17, 18].

The ancient windmills featured basic control mechanisms, often operated manually by millers. Millers would adjust the position of the sails or blades to optimise power generation based on the prevailing wind conditions.

Some windmills employed mechanical governors, albeit simple ones, to regulate the speed of rotation. These windmills had *tail vanes* that could pivot and were used to align the windmill with the prevailing wind direction, ensuring that the blades faced the wind for maximum efficiency.

Tail vanes, also known as *fantails*, were a crucial component of traditional windmills. These devices were mounted on the back of the windmill structure and were designed to pivot or rotate. Their primary purpose was to ensure that the windmill's sails or blades were properly aligned with the prevailing wind direction.

The tail vane was typically a flat or slightly curved panel attached to a pivot point. As the wind changed direction, the vane would respond by pivoting, rotating the entire windmill structure so that the blades would face the wind again.

In many traditional windmills, the operation of the tail vane was manual or semi-manual. The miller or an assistant must monitor the wind direction and adjust the tail vane's position accordingly. This required constant attention, especially in areas with variable wind conditions.

Edmund Lee was an English millwright and inventor who significantly contributed to windmill technology during the eighteenth century. He is particularly known for his improvements to tail vanes and their automatic control mechanisms in 1745 [19]. Lee's inventions aimed to make windmills more efficient and less reliant on constant manual intervention. His innovations included the development of automatic tail vane control systems. These systems used mechanical linkages and gears to connect the tail vane to the main shaft of the windmill. As the wind direction changed, these mechanisms would automatically adjust the position of the tail vane to keep the blades facing the wind. A schematic of this regulatory mechanism is shown in Fig. 2.3.

Fig. 2.3 E. Lee's self-regulating windmill with fantail

This significantly improved the efficiency of windmills by reducing the need for constant manual adjustment. Windmills could operate more reliably and consistently, regardless of changing wind conditions.

Lee's contributions to windmill technology had a lasting impact and are considered important milestones in the development of automated control systems for windmills. His innovations paved the way for more advanced control systems in later wind turbines and windmills generations. His developments in windmill control systems were not isolated achievements but rather part of a larger movement toward increased automation and efficiency in technology.

2.3.2 The Temperature, Flow, Level, and Pressure Regulators

Temperature, flow, level, and pressure regulators played essential roles at the start and during the Industrial Revolution, contributing significantly to that era's rapid technological advancements and industrial transformation.

Early in the Industrial Revolution, basic temperature control mechanisms like manual dampers in furnaces and boilers were used to regulate heat. These elementary controls allowed for some degree of temperature management in industrial processes. Later, the need for precise temperature control became more apparent. Innovations like the thermostat, attributed to Cornelis Drebbel and later improved upon by others, played a pivotal role in various applications, including ovens, furnaces, and incubators.

Flow control systems primarily relied on manual valves and gates to regulate the flow of liquids and gases. These simple mechanisms lacked the precision and automation needed in the emerging technologies. The development of more advanced flow control mechanisms, such as float valves and flowmeters, significantly improved fluid flow regulation. Float regulators, in particular, played a crucial role in maintaining consistent liquid levels in industrial processes like waterwheels and steam boilers.

Basic methods for maintaining liquid levels were replaced with refined float regulators, such as those used in house water reservoirs and steam engines, bringing about more precise and automated level control. These devices ensured a constant water supply for various industrial applications, contributing to the reliability and efficiency of processes.

Pressure regulation was essential for safety in steam boilers and other vessels. Although relatively simple, pressure relief valves helped prevent catastrophic accidents caused by overpressure. As steam engines became central to the Industrial Revolution, more sophisticated pressure regulation systems, including governors, emerged. These devices automatically adjust steam flow to maintain consistent pressure. Pressure regulation also played a key role in developing high-pressure steam engines, increasing their efficiency.

Among the first documented feedback mechanisms for closed-loop control in the pre-classical control era is the remarkable invention by Cornelis Drebbel, a Dutch inventor and engineer who lived in the late 16th and early seventeenth centuries. He created a chicken incubator with a mercury thermostat, automatically regulating its temperature. Drebbel made significant contributions to the development of automatic closed-loop temperature control. He is renowned for inventing the first known *thermostat*, a mercury thermostat. This thermostat consisted of a glass tube filled with mercury connected to a temperature-sensitive bimetallic strip. When the temperature inside an oven or incubator changed, the bimetallic strip would either expand or contract, leading to a corresponding rise or fall in the mercury level. Consequently, this mechanism controlled a damper or ventilation system, enabling the automatic regulation of the oven's temperature.

Drebbel's self-regulating oven had practical applications, including incubating chicken eggs. By maintaining a consistent temperature, his invention improved the success rate of egg hatching. Drebbel's thermostat invention laid the foundation for developing modern temperature control systems. Using a temperature-sensitive element, such as a bimetallic strip or mercury, to regulate heating and cooling systems has evolved and is still a fundamental principle in modern thermostats used for climate control in homes, offices, and industrial settings.

Cornelis Drebbel's contributions have earned him recognition as a pioneering figure in the history of automatic closed-loop control with a particular focus on temperature regulation. A schematic of Drebbel's self-regulating oven, also known as his *thermostat-controlled oven* or *incubator*, is shown in Fig. 2.4, drawn based on Fig. 38 of [20], which is from the original manuscript. In the oven, the fire source is separated from the oven chamber by a metal plate with an opening in the middle.

The oven chamber contains an incubator box with hollow walls filled with water. The flue gasses can escape through the opening at the top of the oven. A temperature sensor is inserted in the water-filled bottom of the incubator. This sensor is a glass …led with alcohol, and its end is a U-shaped open part filled with mercury. With the …nal expansion of the alcohol, the mercury level in the open part rises or falls. If

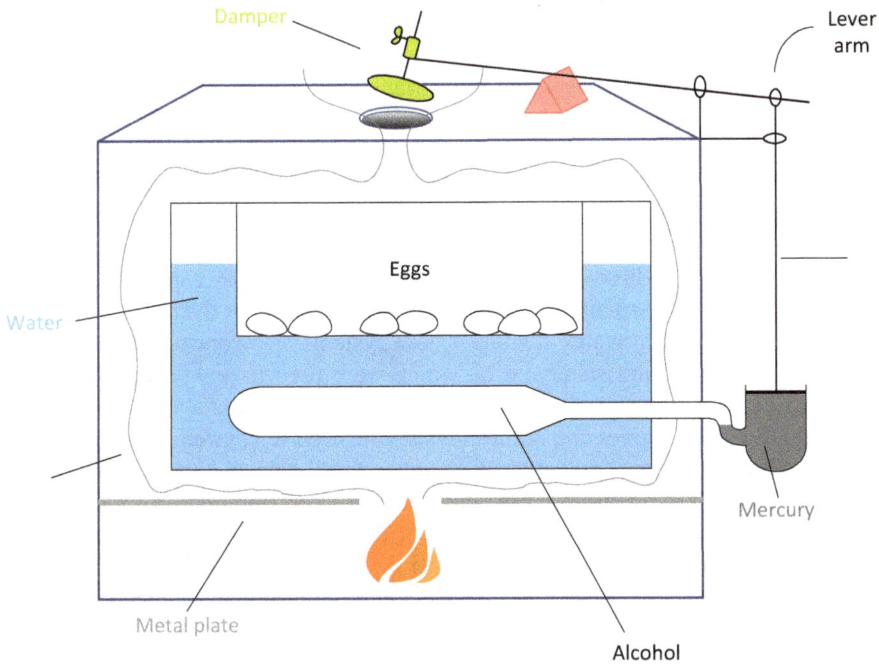

Fig. 2.4 Drebbel's self-regulating oven

the water in the incubator becomes hotter than the desired temperature, the damper will close the flue opening. The fire will then produce less heat by receiving less air.

A reverse action will occur once the water temperature falls below the desired temperature. For more details and a block diagram representation of Drebbel's self-regulating oven, refer to [20]. Both the rod and the lever shown are threated at the ends, thus providing adjustability for the desired temperature or the set-point signal.

This feedback process effectively automates temperature regulation in the oven. It maintains a constant temperature by adjusting airflow in response to internal temperature fluctuations, preventing overheating or excessive cooling.

Drebbel's temperature-regulating feedback system was later developed by other inventors and found many applications in different fields. A notable example is Bonnemain's heat regulation system and hot water network, allowing continuous heating and extended running time. Initially employed in producing goods, the regulator heating system was later adapted for broad applications in space heating systems [21].

Flow and level control systems significantly influenced the Industrial Revolution and marked a profound shift in goods production. These control systems were instrumental in driving technological advancements and the growth of industrial processes. Flow control systems, particularly in the context of steam engines and water-powered machinery, were crucial for regulating the movement of fluids within industrial processes.

Steam engines, at the heart of the Industrial Revolution, relied on precise steam flow control to achieve mechanical work. Controlling the steam flow was vital for optimising engine efficiency and power output. Flow control systems allowed for the controlled movement of materials through pipelines, facilitating processes like the transportation of water, fuels, and raw materials within factories and across industries.

Also, level control systems were essential in managing the quantities of liquids or materials in various industrial processes. In textile mills, for example, level control systems ensured a consistent and even distribution of water or dye, which was critical for producing high-quality textiles. These systems also controlled water levels in canals, reservoirs, and waterwheels, which were used to power machinery and facilitate transportation. The regulation of liquid levels was crucial in the late 1700s, especially in steam engines and domestic water distribution systems. Maintaining a consistent liquid level was essential for steam engines' safe and efficient operation.

In W. Salmon's book from 1746, the prices for ball-and-cock float regulators used for maintaining the level of house water reservoirs are recorded. These early float regulators were used in domestic water systems. Float regulators were found in the first patents for the flush toilet around 1775, with Thomas Crapper, a London plumber, contributing to the refinement of this design [20].

Early use of float valve regulators in steam engines is attributed to J. Brindley's patent in 1758 and S.T. Wood's use in his brewery in 1784. The float regulator became commonly used in steam engines by 1791 when the firm of Boulton and Watt adopted it.

In 1681, Denis Papin invented a safety valve primarily used as a safety device for pressure cookers [20]. While it is believed that he had experimented with his safety valve on steam-driven engines, the widespread adoption of safety valves on steam engines came later during the Industrial Revolution.

The pressure regulator, including safety valves, underwent further refinement during the late 18th and early nineteenth centuries. Engineers like R. Delap and M. Murray contributed to improving pressure regulation technology for steam engines, making them more reliable and efficient.

Boulton and Watt, prominent figures in the development of steam engines during the late eighteenth century, combined a pressure regulator with a float regulator in 1803. This innovation helped maintain constant steam pressure within the engine's boiler, which was crucial for engine performance and safety [20].

2.3.3 *The Centrifugal Governors*

As a key component of early steam engines, *centrifugal governors* played a crucial role in the Industrial Revolution. These governors were critical for maintaining and controlling the speed of steam engines, ensuring their efficiency and safety.

Steam engines were widely used to power various machinery during the Industrial Revolution, from textile mills to factories, to maintain the quality and productivity of industrial processes. Centrifugal governors were used to control the throttle or valve settings of the steam engine, adjusting them automatically to keep the engine running at a predetermined speed.

Maintaining a stable speed was crucial for the safety of steam engines. Uncontrolled increases in engine speed could lead to catastrophic failures and accidents. The centrifugal governor acted as a safety mechanism by preventing the engine from running too fast and potentially causing damage or harm.

Centrifugal governors represent an early example of automation in industrial settings. They could adjust engine settings automatically without needing constant human intervention, allowing for the continuous operation of machinery and reducing the need for manual labour.

The most important component of a centrifugal governor was the pair of spinning balls or weights attached to the governor's arms. These spinning balls are typically referred to as *flyballs*. The flyballs are designed to spin at the end of rotating arms. As the speed of the engine they are regulating changes, the centrifugal force acting on the flyballs causes them to move outward or inward. This movement is directly related to the engine's speed. A schematic of the centrifugal governor is shown in Fig. 2.5.

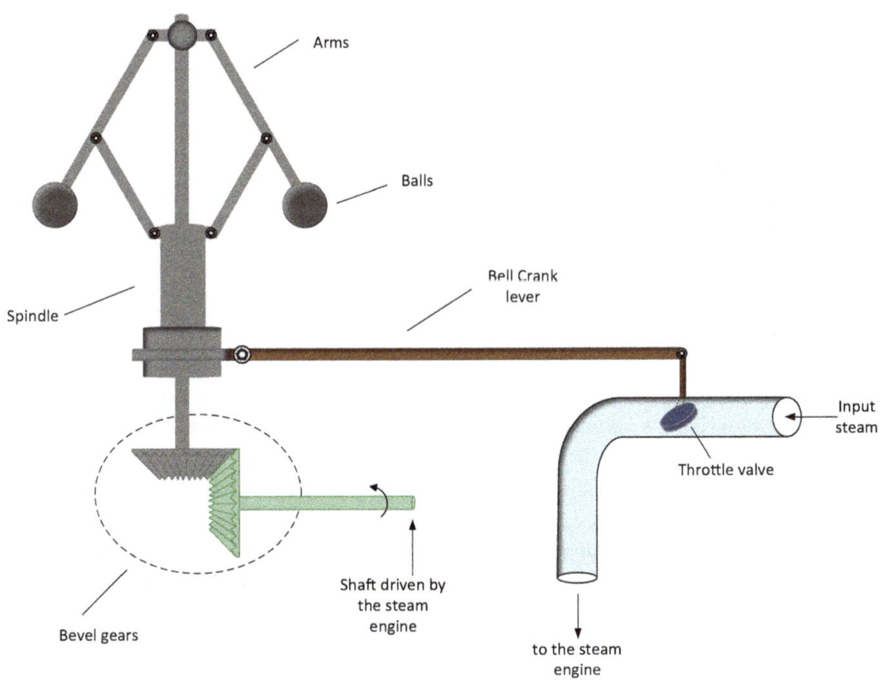

Fig. 2.5 A schematic of the centrifugal governor

When the flyballs move outward due to increased engine speed, they act on a mechanism, such as a linkage or throttle valve, to reduce the amount of steam or fuel supplied to the engine. Conversely, when the engine slows down, the flyballs move inward, causing an increase in the supply of steam or fuel. This dynamic adjustment of engine components allows the governor to *control* the engine's speed.

The movement of the flyballs creates a feedback loop that continuously adjusts the engine's operations to maintain a set speed. This feedback loop ensures that the engine operates at a consistent and desired speed (see Fig. 2.5).

James Watt played a crucial role in developing and improving the centrifugal governor. With its rotary output motion, Watt's steam engine represented a significant technological advancement that reached maturity by 1783. At this point, it had evolved from earlier iterations and had become a more reliable and efficient power source. The primary motivation behind the development of Watt's steam engine was to introduce a prime mover into milling. The fact that the Albion steam mill[1] began operations in early 1786 is evidence of the practicality and impact of Watt's steam engine. This marked the successful application of steam power to milling, revolutionising how grain and other materials were processed. It increased efficiency and had far-reaching economic implications by reducing labour requirements and increasing production capacity.

A key challenge with the centrifugal governor was closed-loop instability, which Watt called *hunting* or *hunting motion*. Hunting was a term used to describe the oscillation or fluctuation in the speed of a steam engine when it was not operating at a constant steady rate. Watt recognised that maintaining a consistent and stable speed in a steam engine was crucial for its efficiency and practical use. The hunting motion was an undesirable phenomenon that occurred when the engine's speed oscillated around the desired set point, and it often resulted from variations in steam pressure, load changes, and the response time of the centrifugal governor.

Watt improved the governor mechanism and other engine components to address this instability. He aimed to minimise the hunting motion and achieve more precise control over the engine's speed. This work led to significant advancements in steam engine technology, making it a more reliable and efficient power source for the industries of his time.

However, as we shall discuss in Sect. 2.4.2, Maxwell's contributions to the stability analysis of Watt's governor provided the effective treatment of hunting. His mathematical and analytical work helped engineers design more stable and effective control systems.

Finally, it is worth mentioning that the flyball governor, with its visible and attractive operation, became an iconic symbol of the new industrial age. Its popularity and widespread recognition contributed to the popular awareness of *feedback control*, making it an essential element of the Industrial Revolution's technological landscape [8].

[1] The Albion Steam Mill was one of the pioneering industrial facilities that used steam power for milling during the Industrial Revolution. It was located in London and is historically significant for being one of the early examples of the application of steam engines to industrial processes.

2.3.4 The First Sampled-Data Systems

The concept and practice of taking samples from a system existed and was applied in various industrial contexts well before the formal theory and sampling techniques were developed and systematically studied. In this section, two early sampled-data systems are introduced. For more details and examples, refer to [22].

The first mechanical watches were developed in Europe during the sixteenth century. These watches were open-loop control systems, operated on a predetermined and consistent mechanism with various gears, springs, and escapements. The accuracy of a mechanical watch was, therefore, solely dependent on the precision of its mechanical components and the quality of its craftsmanship. However, various external factors, such as temperature changes and physical shocks, influenced their accuracy.

In the late eighteenth century, the renowned French-Swiss watchmaker Abraham-Louis Breguet developed a highly sophisticated and luxurious clock and watch set called the *Pendule Sympathique*. The primary purpose of the Pendule Sympathique was to ensure that a pocket watch was always set to the correct time. The clock (pendule) served as a master timekeeper, and the pocket watch was designed to be synchronised with it.

To synchronise the watch with the clock, it was placed in a special cradle within the clock. The clock would then adjust the watch's time, typically using a keyless winding system, ensuring precise timekeeping.

The Pendule Sympathique was a specialised device for speed regulation or synchronisation. There was a mount within the clock for placing a pocket watch. The pocket watch was placed on the clock before 12 o'clock. Then, a pin emerged from the clock and automatically adjusted the watch's balance spring regulating arm. This transmitted the accuracy of the clock to the smaller portable pocket watch.

A high level of synchronisation was attainable within a relatively short period, typically two or three days. This observation underscores the efficiency of the *sampling* scheme used. By periodically sampling and adjusting, the system can rapidly reach a state of synchronisation, ensuring that the timekeeping accuracy is maintained.

Breguet created a limited number of Pendule Sympathique sets, now considered some of the most exceptional examples of horological artistry and innovation from the late 18th and early nineteenth centuries. For additional details on Breguet's automatic synchronising mechanism, including photos and a block diagram representation, refer to [20].

Another notable example of a sampled-data mechanism is the *Gouy Regulator*, depicted in Fig. 2.6 [22]. It is a *thermostat* or *temperature control* mechanism developed by Gabriel Léon Gouy in 1897. Its primary goal was to ensure that the oven's temperature remained constant regardless of external disturbances.

The Gouy Regulator employs an electrical contact rod immersed in a mercury pool. When the electrical contact rod contacts the mercury, it forms an electrical circuit, allowing current to flow through the relay coil. The current flowing through

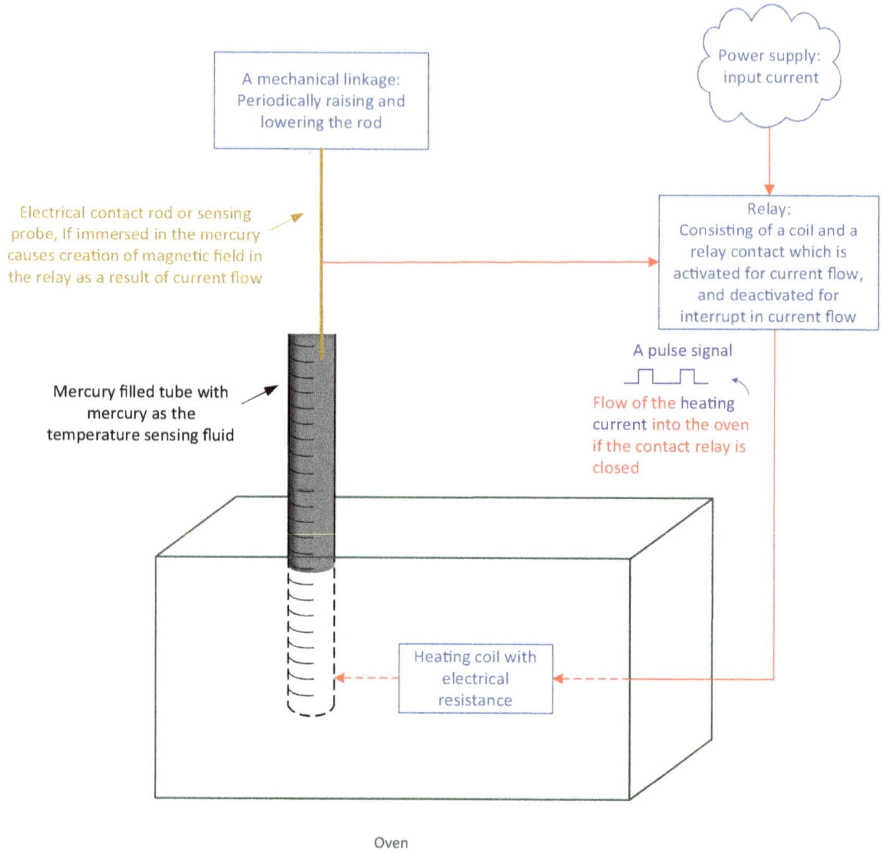

Fig. 2.6 The Gouy regulator

the relay coil directly influences the relay's operation. As the relay coil receives current, it activates the relay, causing it to switch from its normally closed state to an open state. When the relay opens, it interrupts the flow of heating current to the oven, effectively turning off or reducing the heat source.

The level of the mercury pool is temperature-dependent. It rises and falls with changes in the oven's temperature. As the temperature increases, the mercury level in the control chamber rises, bringing the electrical contact rod into contact with the mercury, completing the circuit, and activating the relay. Conversely, the mercury level drops as the temperature decreases, breaking the circuit and deactivating the relay.

In this system, the control of the heating element is not achieved by varying the amplitude of the heating current but by varying the duration or width of pulses. When the relay opens and interrupts the heating current, it periodically creates a series of pulses. The width or duration of these pulses varies in proportion to the oven's temperature. Higher temperatures result in wider pulses, while lower temperatures result in narrower pulses. The pulses are all of equal amplitude, meaning the strength of the heating current remains constant; only the on–off timing (duty cycle) changes.

The system described, where the temperature control is achieved by varying the pulse widths, is a form of *pulse-width modulation* (PWM) *feedback control*. In PWM control, the signal information is encoded in the width of the pulses. Wider pulses represent higher temperatures, and narrower pulses represent lower temperatures. PWM systems are known for their precision in controlling systems like ovens, where maintaining a consistent temperature is crucial.

A key point is that the control system's design achievements presented in this section are not based on mathematical principles. Mathematical modelling and rigour were not utilised in their design. The next section discusses the integration of mathematics into control system analysis and design.

2.4 Emergence of Mathematical Principles in Control Theory

As we understand it today, control theory owes its existence and evolution to the integration of mathematical principles. At its core, control theory deals with the challenge of governing dynamic systems, ensuring they respond appropriately to input signals or disturbances as defined by the designer. As shown in the previous sections, this quest for control has deep historical roots, but it was not until the late 19th and early twentieth centuries that *mathematical thinking* and *tools* took centre stage, enabling engineers and scientists to address problems with unprecedented precision. *Mathematical rigour* replaced the need for *skilled craftsmanship* and *meticulous workmanship*. This section will briefly explore some fundamental mathematical concepts that gave birth to mathematical control theory.

2.4.1 The Foundations: Calculus of Variations and Optimal Control

Calculus of variations is a mathematical field that addresses fundamental questions of optimisation. The calculus of variations laid the groundwork for optimal control theory, offering a framework to determine the paths, or control strategies, that minimise or maximise specific performance criteria. This revolutionary concept led to the *notion of optimality, which* became a central theme in control systems design.

As is stated in [23], "Optimal control was born in 1697—300 years ago—in Groningen, a university town in the north of The Netherlands, when Johann Bernoulli, professor of mathematics at the local university from 1695 to 1705, published his solution of the brachystochrone problem." Bernoulli posed the brachystochrone problem in the June 1696 issue of *Acta Eurditorum*.[2] It is the problem of finding the curve of the quickest descent between two given points in a vertical plane. This fascinating problem, which is truly the ancestor of present-day optimal control and optimisation problems, is reported in [23] as Johann Bernoulli's announcement:

> **Invitation to all mathematicians to solve a new problem**
>
> If in a vertical plane two points A and B are given, then it is required to specify the orbit AMB of the moveable point M, along which it, starting from A, and the influence of its own weight, arrives at B in the shortest possible time. So that those who are keen of such matters will be tempted to solve this problem, it is good to know that it is not, as it may seem, purely speculative and without practical use. Rather it even appears, and this may be hard to believe, that it is very useful also for other branches of science than mechanics. In order to avoid a hasty conclusion, it should be remarked that the straight line is not the one which is travelled in the shortest time. However, the curve AMB—which I shall divulge if by the end of this year nobody else has found it—is very well known among geometers.

For some earlier optimisation problems, refer to [23]. However, it must be emphasised that the brachistochrone problem is the first mathematically challenging and groundbreaking optimisation problem that attracted the attention of many prominent mathematicians. Several eminent mathematicians, including Johann and his elder brother Jakob Bernoulli, Leibnitz, Tschirnhaus, L'Hôpital, and Newton, provided solutions to the problem. For details, refer to [23].

These solutions were remarkable and initiated a period of intensive research on related problems. A notable example is the work of Euler and Lagrange, which led to the well-known results in the calculus of variations. It is believed that the papers published in response to the brachistochrone problem provided the proper platform for the emergence of the field of calculus of variations.

Later, in 1744, Euler published a book that addressed the maximum and minimum of plane curves, deriving what later became known as Euler's equations. While studying the same problem, Lagrange provided a more accessible solution and derived a necessary condition known as the *Euler–Lagrange equation* in the 1750s and 1760s. It is desired to minimise the following functional or cost function

$$J = \int_a^b L(q(t), \dot{q}(t), t)dt$$

where $L(\cdot)$ is a given function. The general form of the Euler–Lagrange equation is

$$\frac{d}{dt}\frac{\partial L}{\partial \dot{q}} = \frac{\partial L}{\partial q}$$

[2] *Acta Eruditorum* was one of the first and most prominent scholarly journals in Europe during the late 17th and early eighteenth centuries. It played a significant role in the dissemination of scientific and academic knowledge during the Enlightenment period. It covered a wide range of topics, including mathematics, physics, astronomy, philosophy, history, linguistics, and literature.

This condition only provides conditions for stationarity, ensuring that the first variation of the functional is zero. However, it did not provide sufficient information to confirm that a function minimises the function. Legendre took the next step to address this. Legendre considered the second variation, which is the variation of the functional with respect to small perturbations of the function $q(t)$. He showed that if the second derivative of $L(\cdot)$ with respect to the velocity term $\dot{q}(t)$ is non-negative, then the function $q(t)$ is indeed a minimum. That is,

$$\frac{\partial^2 L}{\partial \dot{q}^2}(q(t), \dot{q}(t), t) \geq 0$$

This result, which was presented to the French Academy of Sciences in 1786, provided an additional condition for the minimisation of functionals and further advanced the calculus of variations.

Later, Karl Weierstrass, a 19th-century mathematician known for his work in real analysis and mathematical rigour contributed to the field of calculus of variations by providing a more rigorous foundation for the results of Euler–Lagrange and Legendre. Weierstrass's work helped clarify and solidify the mathematical foundations of the calculus of variations, ensuring that solutions to variational problems were valid extrema. While Weierstrass's work may not be directly tied to optimisation and optimal control theory that was later developed, the broader mathematical principles and rigour that he advocated have shaped the mathematical frameworks and techniques used in optimisation and optimal control theory.

2.4.2 The Stability Problem

The English adjective *stable* originates from the Latin *stabilis*, derived from *stare*, meaning *to stand*. Its primary meaning is *standing firmly* or *firmly established*. A natural extension of this concept is *durable*, not to mention the moral aspect of being *steady in purpose* or *constant*. As it stands, this concept of stability appears clear and useful in everyday life. However, for utilisation in engineering systems, the stability concept needed to be refined and defined with more technical details.

In the early stages, the concept of stability was specialised in mechanics to describe a specific type of equilibrium in a material particle or system. The equilibrium is considered stable if, following any sufficiently small perturbations, the movements remain arbitrarily close to it, with an arbitrarily small velocity.

While this mechanical definition of stability, formulated in precise mathematical terms, proved useful in many situations, it was inadequate in others. Consequently, over the years, numerous other concepts have been introduced. Each concept is more or less related to the initial definition and the common-sense meaning of stability and was created for specific technical or physical purposes [24].

Before the development of the mathematical stability tests commonly used in control systems theory today, stability in control systems during the Industrial Revolution and earlier periods was often inferred through empirical observation, trial and error, and practical engineering experience. Engineers and control system operators relied on their *intuition* and *practical knowledge* to design and maintain stable control systems. The key methods and principles they used can be summarised as *Trial and error*, *Heuristic rules*, some type of *Physical prototyping*, *Knowledge of experienced engineers*, and *Empirical observations*.

It is important to note that while these methods were effective to some extent, they were limited by the complexity of the systems and the lack of a systematic mathematical framework for analysing and ensuring closed-loop stability. These methods did not provide a general platform for stability analysis, and it was therefore necessary to start each system from scratch.

The mathematical techniques of calculus of variations outlined in Sect. 2.4.1 did not directly affect the trend of control systems engineers when they were presented. These results were acknowledged in control engineering in the twentieth century with the introduction of the notion of optimality. It can be argued that mathematics formally entered the realm of control theory and design with the advent of models for the first time in the seminal paper of Maxwell *On Governers* [25]. Norbert Wiener, in introducing the word *cybernetics* in [26] describes the Mawell paper as "… the first significant paper on feedback mechanisms is an article on governors, which was published by Clerk Maxwell in 1868" and in [27] Maxwell is recognised as the "father of control theory".

The Maxwell magic was to introduce differential equations in modelling the behaviour of the flyball governor feedback control system invented by James Watt in 1788. Maxwell's ground-breaking contribution introduced mathematical modelling in the *stability analysis* of a closed-loop control system. An idea that soon found many applications and advocates and solved many unsolved stability analysis problems.

The differential equations encountered in the flyball governor model were nonlinear. By linearising these nonlinear equations, Maxwell introduced the notions of what is today called *real poles*, *imaginary poles,* and the significance of the *pole position* in the right half plane. This model-based approach to the stability analysis of a control system through the differential equations of motion was performed for the first time in the history of control theory. Hence, it is plausible to introduce Maxwell as the pioneer of the model-based control theory, particularly mathematical stability analysis.

Maxwell describes in [25] that the *motion* of a machine with its governers can fall into one of the four categories: *continually increasing*, *continually diminishing*, an *oscillation of continually increasing amplitude*, and an *oscillation of continually decreasing amplitude.* He then asserts that "the first and third cases are evidently inconsistent with the stability of the motion, and the second and fourth alone are admissible in a good governor. This condition is mathematically equivalent to the condition that all the possible roots, and all the possible parts of the impossible[3]

[3] Imaginary as we call it today.

roots, of a certain equation[4] shall be negative." He then confers that "I have not been able completely to determine these conditions for equations of a higher degree than the third; but I hope that the subject will obtain the attention of mathematicians."

Maxwell's call for mathematicians to tackle the stability problem in a general case was welcomed by mathematicians and the subject of *Stability of Motion* became a topic for the Adams Prize[5] essay competition in 1858.

Edward John Routh proposed a solution in his famous Adams Prize essay, where he developed a stability table or algorithm that has not surpassed its computational simplicity. His essay on stability won him the Adams Prize in 1860. It is interesting that Maxwell was one of the examiners for the Adams Prize and was very well aware of the practical importance of Routh's mathematical achievement.

Routh came to the stability problem from practical considerations. However, some mathematicians have tackled the same problem from a completely theoretical perspective.

Charles Hermite, a renowned French mathematician known for his significant contributions to various mathematical fields, was primarily interested in establishing criteria for roots with positive imaginary parts. Additionally, Hermite's work laid the foundation for utilising quadratic forms in stability investigations. Today, we recognise that quadratic forms offer a robust mathematical framework for the analysis, design, and optimisation of control systems.

Hermite's emphasis on positive imaginary roots, as opposed to negative imaginary roots, may have been influenced by the simplicity of mathematical formulation. Importantly, Hermite's methodology extends to determining the distribution of roots in the complex plane using linear fractional transformations. His contributions significantly aided Hurwitz in developing Hurwitz's celebrated criterion for left-half-plane stability, which establishes necessary and sufficient conditions for the roots of an algebraic equation to have negative real parts.

A German-Swiss mathematician, Hurwitz authored his groundbreaking paper in 1895 while a professor at the ETH[6] in Zurich. The impetus for his research on the stability problem stemmed from his colleague Stodola[7] at ETH. Similar to Routh's work, the motivation behind this research was rooted in practical considerations, specifically the design of turbines.

---

[4] Characteristic equation as we call it today.

[5] The Adams Prize is a prestigious mathematics prize established at the University of Cambridge in 1848 by Sir George Biddell Airy. It is awarded for the best essay on a specified subject in mathematics or astronomy.

[6] ETH Zurich, often referred to simply as ETH, stands for Eidgenössische Technische Hochschule Zürich in German, which translates to the Swiss Federal Institute of Technology Zurich in English.

[7] Aurel Stodola (1859–1942) was a prominent Swiss engineer and physicist of Slovak origin. He is best known for his significant contributions to the field of thermodynamics and fluid dynamics, particularly in the context of steam turbines. Stodola made pioneering advancements in the design and analysis of steam turbines. He provided the motivation for research on the stability problem related to turbines at the ETH in Zurich. His practical engineering concerns and collaboration with mathematicians like Adolf Hurwitz contributed to advancements in stability analysis in control theory.

Remarkably, Hurwitz was not aware of Routh's earlier works. His results, however, are closely related to the work of Hermite. In those days, scientific communication faced significant challenges. Consequently, the criterion later became known as the *Routh-Hurwitz criterion.* For further details, refer to [28].

The methodology developed by Hurwitz applied to characteristic equations with real and complex coefficients. The Hurwitz stability criterion involved a specific sequence of positive determinants, now called *Hurwitz determinants.* It was applied to polynomials with real coefficients, which had roots with negative real parts, known as *Hurwitz polynomials.*

This pioneering criterion found practical application in 1894 during the Davos Spa Turbine Plant design in Switzerland. Remarkably, this is considered one of the earliest instances where stability conditions were employed in the practical design of a control system [29].

2.5 Russian and Soviet Contributions

An excellent account of the Russian and Soviet contributions to the development of control engineering can be found in [30]. In this section, drawing from [30], we will specifically focus on the contributions of Russian and Soviet scientists in stability theory.

A significant challenge in the latter half of the nineteenth century that fueled the interest in control theory and stability studies in Russia and Western Europe was the regulation of steam engines. In Russia, a noteworthy example is the 1871 publication titled *On The Centrifugal Governor* by the renowned mathematician P. L. Chebyshev. This work runs in parallel with the efforts of Maxwell, as discussed in Sect. 2.4.2. The problem of hunting and instability in hundreds of thousands of engines resulted in considerable economic costs. During that period, many scientists and engineers attempted to address the instability of these systems by modelling them and deriving stability conditions. Maxwell and Routh's pioneering and seminal work marked the first theoretical solution to the stability problem in Britain. However, their results remained relatively unknown to other parts of the world for quite some time. Initially, their findings had limited impact on industrial control problems. Engineers, if they were aware of the work at all, often found the assumptions about the nature of governors unrealistic and the mathematical aspects inaccessible.

The first Russian to tackle the same problem took a very different approach. Ivan Alekseevich Vyshnegradskii, a practical engineer and professor at St. Petersburg's engineering school, published a paper in Russian in 1877 addressing the stability of steam engines. In his paper, Vyshnegradskii omitted consideration of Coulomb friction and modelled the engine as an integrator and the governor as a second-order system by linearising the system around an operating point. By an ingenious change of variables, he transformed the third-order characteristic equation into

$$\varnothing^3 + x\varnothing^2 + y\varnothing + 1 = 0$$

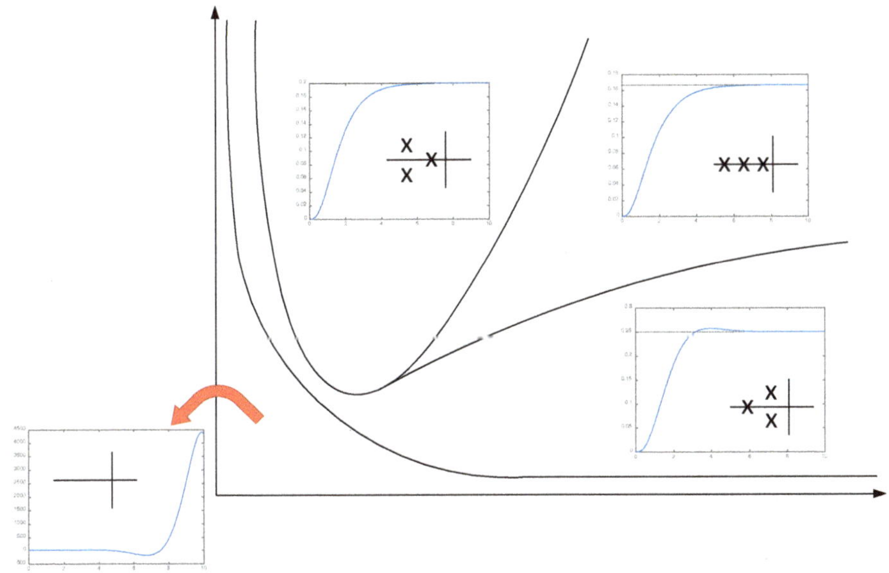

Fig. 2.7 Vyshnegradskii stability diagram. The corresponding pole locations and system step responses

where the parameters x and y depend on system characteristics. These parameters became known in the Russian and German literature as the *Vyshnegradskii parameters*. Vyshnegradskii demonstrated that the roots' location in this equation determines the overall shape of the system's transient response. Figure 2.7 depicts the relationship between root location and the system's transient response, as derived by Vyshnegradskii.

Comparing Vyshnegradskii's results with Maxwell's, we can conclude the following:

- Vyshnegradskii's results were more practically oriented and easier to apply.
- Maxwell's results were more theoretically tractable.
- Vyshnegradskii's results had a far-reaching influence both inside and outside Russia.

In fact, Vyshnegradskii's findings served as the foundation for the work of Stodola and Hurwitz, and his famous diagram continued to be a fundamental component in Soviet control texts until the 1960s. Vyshnegradskii's analysis, particularly in light of Stodola and Hurwitz's subsequent treatment of higher-order linear systems, provided continental European engineers with an accessible approach to machine regulation. It was incorporated into influential textbooks on governors. For further details, refer to [30].

The Lyapunov's Legacy. On 12 October 1892, Alexandr Mikhailovich Lyapunov defended his doctoral thesis titled *The general problem of the stability of motion* at Moscow University. The results of this thesis have since become one of the most widely utilised research achievements of modern times. It is stated in [30] that "Rarely can a doctoral thesis have had such an enormous influence on a scientific or engineering discipline." In 1992, marking the centenary of its publication, an issue of the International Journal of Control was dedicated to Lyapunov's work. The issue included an English translation of the 1892 memoir, a bibliography of Lyapunov's published works, and a biography translation [31]. Additionally, many papers on Lyapunov's work were published, for example, refer to [32, 30]. Lyapunov is best described by the inscription on his gravestone as *The creator of the theory of stability of motion, the doctrine of equilibrium figures of rotating liquid, methods of the qualitative theory of differential equations, author of the central limit theorem and other deep investigations in areas of mechanics and mathematical analysis.*

In his thesis, Lyapunov presented two main methods for investigating the stability of motion. Lyapunov's *first* method provided theoretical validity for the commonly employed technique of linearising a nonlinear system and studying its behaviour at equilibrium points. Lyapunov's *second* or *direct* method introduces generalised energy functions or what we now call *Lyapunov functions* and infers general nonlinear system stability by studying these functions and their derivatives. Lyapunov succeeded in demonstrating the asymptotic stability of a nonlinear system without the necessity of finding explicit solutions to its nonlinear differential equation. The objective of the Lyapunov' second method is stated in the words of Kalman "*To answer questions of stability of differential equations, utilizing the given form of the equations but without explicit knowledge of the solutions.*" Where he calls it "*One of the most important events in the theory of stability of dynamic systems*" [33].

These compelling ideas on stability emerged in Lyapunov's mind, influenced by his understanding of astronomical problems and the foundational work of Lagrange, Laplace, Dirichlet, and Liouville. In astronomical problems, *stability* or *Lyapunov stability* was preferred rather than asymptotic stability. The intensified attention to asymptotic stability, characteristic of contemporary control theory, did not materialise until the twentieth century.

Lyapunov's second method states that for a given dynamical system, the stability of an equilibrium can be established if a positive-definite function of the state coordinates exists, referred to as the generalised energy function. This function should exhibit a rate of change, following a small displacement from equilibrium, that is consistently negative or zero. In what follows, a brief historical background of this seminal result is given from [31].

In the eighteenth century, astronomers and mathematicians dedicated significant efforts to demonstrate that the observed deviations of planets and satellites from fixed elliptical orbits align with Newton's principle of universal gravitation. This alignment, however, required a thorough consideration of the disturbing forces exerted by celestial bodies on each other. These deviations manifested in two forms: firstly, oscillatory motions with relatively short periods (on the order of a few years), and secondly, residual slow changes in the ellipse parameters. These changes could be non-oscillatory or have very long periods, potentially spanning tens of thousands of years. The first type, known as *periodic inequalities*, could be explained as a body's response to the periodic forces exerted by its neighbouring bodies as they continuously trace their orbits. The second type, *secular inequalities*, raised a crucial question for the solar system: would these secular inequalities accumulate over millennia and jeopardise the system's stability?

Laplace investigated secular inequalities in the motion of satellites in a system. He considered a system of k satellites, each moving in an elliptical orbit around a central body with a relatively large mass. The parameters describing each satellite's orbit included the major semi-axis (a_j), eccentricity (e_j), and inclination angle (i_j) of the ellipse. Each satellite, denoted as the jth satellite, has its mass (m_j), a_j, e_j, and i_j. Laplace and Lagrange had found, through earlier work, that the satellite orbits a_j are at least approximately not subject to secular inequalities. This meant that the size of the ellipses (a_j) remained relatively constant over long periods in their analysis. This property allowed Laplace to treat them as constants in his calculations. This simplified the analysis, making it more manageable. While treating the major semi-axes as constants, Laplace allowed e_j and inclination angle i_j of the satellite orbits to vary with time. These variations were the focus of Laplace's investigation. Laplace restricted attention to secular inequalities, studying the long-term variations in eccentricity and inclination of the satellite orbits. Gravitational interactions and perturbations likely influence these variations within the system. Laplace claimed that

$$V_1\left(e_1\ e_2\ \cdots\ e_k\right) \equiv \sum_{j=1}^{k} m_j\left(a_j\right)^{1/2} e_j^2 = constant$$

$$V_2\left(i_1\ i_2\ \cdots\ i_k\right) \equiv \sum_{j=1}^{k} m_j\left(a_j\right)^{1/2} \tan^2 i_j = constant$$

These expressions represented contributions to the total angular momentum made by eccentricities and inclinations. The assumption was that these parameters were small. The constancy of these contributions was related to the conservation of angular momentum, a key principle in celestial mechanics. Note that $tan i_j$ could be replaced by i_j, assuming small inclinations. Laplace's analysis suggested that if the eccentricities and inclinations were initially small, they would remain small over time, implying stability in the solar system. The constancy of these expressions indicated that these parameters did not grow uncontrollably, supporting the hypothesis of solar system stability.

However, Laplace's conclusion about stability was not rigorously established due to various approximations made in the analysis. While the results suggest stability, the analysis may not have accounted for all complexities with absolute precision. Laplace defined the functions V_1 and V_2, which are positive-definite and have zero derivatives. These functions align with what we now refer to as Lyapunov functions. Lyapunov, the son of an astronomer, was primarily engaged in mathematical challenges within astronomy. Consequently, he would have had acquaintance with Laplace's utilisation of V_1 and V_2 in the investigation of stability.

Another part of the background against which Lyapunov's technique was developed is Torricelli's principle, which was already used in the seventeenth century. The foundation of Torricelli's principle asserted that in an equilibrium state, a system of interconnected heavy bodies positions its centre of gravity at the lowest point. A contemporary interpretation extends this concept to state that, in a mechanical system, an arrangement achieving minimum potential energy represents a point of equilibrium.

In 1788, Lagrange proposed that a stable equilibrium point corresponds to an isolated minimum of potential energy within a conservative system, characterised by energy conservation. He excluded nonlinear terms from the differential equations in his proof and used a simplified linear system that may not fully represent the complexities of the real world. Dirichlet, in 1846, attempted to overcome the limitations of Lagrange's proof by offering a more rigorous demonstration of the principle.

Joseph Liouville employed Lagrange's principle in examining equilibrium figures of rotating fluid bodies, where gravitational attraction exists between the particles. Notably, Liouville utilised a principle based on maximum kinetic energy rather than minimum potential energy. However, these two principles are equivalent due to the constancy of total energy.

In 1882, Chebyshev presented Lyapunov with a challenging problem: to investigate whether non-ellipsoidal figures of equilibrium for rotating fluid bodies could exist in proximity to the well-known ellipsoidal figures. This inquiry prompted Lyapunov to study the stability of ellipsoidal figures of equilibrium, leading to the publication of his master's dissertation on the subject in 1884. Following a similar approach to Liouville, Lyapunov examined conditions for the figure to possess minimum potential energy, considering modifications to account for centrifugal force. Presumably, Lyapunov became acquainted with Liouville's work on rotating fluids through Chebyshev, who frequently traveled to France and maintained a friendship with Liouville.

Approximately eight years before publishing his 1892 memoir on stability, which served as his doctoral dissertation, Lyapunov thoroughly understood Lagrange's principle of minimum potential energy. It likely struck him that in Dirichlet's proof, the essential aspect was not solely that the total energy is composed of two positive-definite functions. Instead, what mattered was the essential nature of the total energy being positive-definite and having zero rate of change—essentially what we would now describe as the total energy being a Lyapunov function.

Lyapunov's (1892) thesis contained many mathematical examples without directly involving engineering stability problems. Application of Lyapunov's stability theory in the control systems analysis and design came many years later, notably in the work of N. G. Chetayev and I. G. Malkin at the Kazan Aviation Institute in the 1930s [32]. Chetaev applied Lyapunov's second method to address aeronautical stability issues, as illustrated in a classic scenario involving the spin stabilisation of a rocket or artillery shell.

Lyapunov's stability theory found its way into control design problems in the Western world during the late 1950s and early 1960s. The fact that Lyapunov's findings remained largely undiscovered by Western control engineers is a classic illustration of the communication gaps within the technology domain.

The recognition of Lyapunov's theory experienced a significant boost due to the complex control challenges arising from the space race between the USA and the USSR. The launch of the Soviet Sputnik 1 initiated this competition.[8] The increased recognition of Lyapunov's second method beyond Russia can be attributed, in large part, to significant events such as the 1st World Congress of the International Federation of Automatic Control[9] (IFAC) held in Moscow in June 1960. Additionally, a pivotal article by R. E. Kalman and J. E. Bertram published in the same month [33], played a crucial role. Notably, a group led by Solomon Lefschetz at the Research Institute for Advanced Study in Baltimore, USA, which included members R. E. Kalman and J. P. La Salle, actively contributed to circulating Russian mathematical techniques during this period. Furthermore, the initiation of cover-to-cover English translations of selected Russian journals in these years significantly contributed to broader awareness. Also, the book by Solomon Lefschetz and John LaSalle, *Stability by Liapunov's Direct Method with Applications*, published in 1961 [34], is one of the early significant works on Lyapunov stability theory [32].

2.6 Pre-classical Control Era: The First Decades of the Twentieth Century

This period of control systems history is very well documented in [35] and is dealt with in [36, 37]. This section is based on the materials from [36].

2.6.1 Innovation, Challenges, and Foundation for the Future

Innovations. Towards the end of the early twentiethcentury, control system engineering was a boosting profession with a large market and a noticeable increase in

[8] Sputnik 1 was the first artificial Earth satellite and a major milestone in the space age. It was launched by the Soviet Union on October 4, 1957. Sputnik 1 marked the beginning of human-made objects orbiting the Earth and the start of the space race between the United States and the Soviet Union during the Cold War. Sputnik 1 was a spherical satellite equipped with four external radio antennas to broadcast radio pulses. Its successful launch had significant scientific, political, and military implications, and it had a profound impact on the course of space exploration history. The event triggered increased interest and investment in space exploration by both superpowers, leading to subsequent developments such as the formation of NASA in the United States and the advancement of space technologies.

[9] IFAC is a global organization that was established in September 1957 and focuses on the study of automatic control and its application in engineering, science, and the influence of control technology on society.

the sales of controllers. This period saw the recognition of the importance of automated control systems in various industries. A general classification of the application of the innovative control systems can be given as follows:

Voltage, Current, and Frequency Regulation: Feedback controllers played a crucial role in regulating voltage, current, and frequency. For instance, automatic voltage regulators (AVRs) were developed to maintain a consistent voltage level in electrical grids despite fluctuations in load. This ensured a stable power supply, especially in rapidly industrialising regions.

Boiler Control for Steam Generation: Boiler control systems became increasingly important in industries reliant on steam power. Controllers were used to control the steam pressure and factors like water level and combustion efficiency.

Electric Motor Speed Control: The advent of electric motors brought about a need for speed control. Controllers, such as early versions of variable frequency drives (VFDs), allowed for precise control of motor speeds.

Ship and Aircraft Steering and Auto Stabilisation: Automated control systems became vital for ship and aircraft navigation. These systems included controllers for steering and auto-stabilisation.

Temperature, Pressure, and Flow Control in Process Industries: Process industries, such as chemical and petrochemical plants, required precise control over temperature, pressure, and flow rates to ensure the safety and efficiency of their operations. Controllers were employed to maintain these parameters within specified ranges.

PID Control: PID control was a significant development, allowing for effective regulation of systems by adjusting control variables based on proportional, integral, and derivative terms.

The emergence of controllers during the early twentieth century represented a critical step in the industrialisation and automation of various processes ranging from electricity generation to manufacturing and transportation.

Challenges. Parallel to the exponentially increasing need for effective controllers in a wide range of industries, the control systems engineering community faced serious challenges in designing controllers during the early twentieth century. These challenges included:

Limited Understanding of System Dynamics: One of the primary challenges faced during this era was the limited understanding of system dynamics. Engineers and scientists were still in the early stages of comprehending how various industrial systems behaved. Designing effective control systems was a formidable task without a clear understanding of these dynamics. For example, in the context of steam boilers, understanding the intricate interplay between water level, steam pressure, and combustion dynamics was essential for efficient control.

Measuring and Actuating Device Limitations: The instruments available at the time were not very accurate or reliable. This could lead to imprecise control and the need for frequent manual adjustments. For instance, early pressure gauges or temperature sensors might not have provided the accuracy required for precise control.

Complex Control Laws: In certain applications, especially those involving complex mechanisms, designing control laws posed a significant challenge. For example, the automatic ship-steering mechanism developed by Elmer Sperry, which incorporated PID control and automatic gain adjustment, represented a sophisticated control system for its time. Designing and fine-tuning such control laws required both expertise and experimentation.

Changing Environmental Conditions: Some control systems, such as those used in marine and aviation applications, had to contend with changing environmental conditions. This required the development of control algorithms capable of adjusting to environmental disturbances and changes.

Lack of Computing Power: During this period, computing power was extremely limited. Designing and implementing control algorithms posed a significant challenge. Engineers often had to rely on manual calculations and physical experimentation to develop control systems.

Limited Data and Communication: Data collection and communication technologies were undeveloped at best. Gathering real-time data from various system parts and using it for control was challenging. Designing closed-loop control systems was a formidable task without efficient data collection and communication.

Simple Regulation versus Complex Control: Many applications were primarily concerned with simple regulation, where basic control strategies sufficed. However, as technology advanced, more complex control challenges emerged, requiring a deeper understanding of control theory and system dynamics.

Foundations for the future. The above challenges had several important implications for developing control engineering as a discipline. Control engineers of that era often had to rely on empirical methods and practical experience. With limited theoretical knowledge and computing power, they had to design and tune control systems through trial and error. This empirical approach led to valuable practical experiences but resulted in less efficient or optimised control solutions with no deep insight into the internal dynamical behaviours or mathematically describable interactions of the different closed-loop system parts.

Engineers needed hands-on expertise in the specific systems they were controlling. Understanding the intricacies of boilers, electric motors, ships, aircraft, or other industrial processes was essential for effective control system design and maintenance.

The challenges encouraged innovation and experimentation. Engineers had to invent new control strategies and mechanisms to address the limitations of existing technology. However, as applications multiplied, engineers faced the problem of the designed controllers that worked satisfactorily for one application or set of conditions performing poorly when applied to different systems or conditions.

The lack of insights led to problems. A change in one part of the system under control, including the process, controller, sensors, or actuator, could change the major time constants. This caused poor closed-loop performances or even instabilities in an apparently well-designed control system.

Control engineers had to possess interdisciplinary knowledge. They needed to understand not only control theory but also the physics, chemistry, or engineering principles of the systems they were controlling.

Note that the lack of theoretical understanding of dynamic systems with no available common mathematical language to discuss the problems and engage with other control engineers in other disciplines made control systems design a difficult multidisciplinary task. Another notable feature of this era was that no simple, easily applied analysis and design methods existed. The only available analysis tool for the control engineers was the differential equation description of their respective systems and the not widely known stability criteria, such as the Routh-Hurwitz stability test. The control engineer had to obtain parameter values for closed-loop stability with no general insight into the degree of stability or the stabilising parameters.

Despite the challenges and limitations, the work of control engineers during this period laid the foundation for future advances in control systems. Their pioneering efforts in developing feedback controllers and control strategies paved the way for today's more sophisticated and automated control systems.

2.6.2 Innovators Elmer Sperry and Nicolas Minorsky

One notable control engineer from the early twentieth century is Elmer Ambrose Sperry. He was an American inventor and engineer known for his pioneering work in control systems, particularly in automatic ship steering. Sperry is perhaps most famous for developing the *gyroscopic compass* and the *gyroscopic stabiliser*. His gyroscopic compass, patented in 1908, provided a stable and accurate means of determining direction for ships, reducing the reliance on magnetic compasses that were prone to interference.

Moreover, Sperry's development of the gyroscopic stabiliser allowed ships to maintain stability even in rough seas and revolutionised ship control systems.

Another noteworthy invention associated with Sperry is the Sperry Automatic Pilot, which utilised feedback control mechanisms to steer ships automatically. This device incorporated advanced control techniques, including proportional, integral, and derivative (PID) control, to adjust the ship's course in response to external disturbances such as wind and waves.

Sperry's gyroscopic stabiliser and automatic pilots laid the foundation for similar systems used in aircraft. These innovations significantly improved the stability and control of aircraft, especially during adverse weather conditions. Gyroscopic technology became an essential component in aviation instrumentation and autopilot systems.

Another notable control engineer from the early twentieth century is Nicolas Minorsky. He was a Russian-American engineer who significantly contributed to the development of the PID control system. Minorsky is widely recognised for his work on automatic steering systems for ships. In 1922, he published a seminal paper titled *Directional Stability of Automatically Steered Bodies* [38]. In this paper, he introduced the concept of using a control system with proportional, integral, and derivative components to maintain the stability of a ship's course. This work laid the theoretical foundation for the PID controllers.

2.6.3 The Birth of Negative Feedback Amplifiers

A telephone repeater, also known simply as a *repeater*, was a device used in telecommunications to amplify and regenerate electrical signals transmitted over long distances in telephone lines or communication networks. Its primary purpose was to compensate for the signal loss as a telephone signal travels over extended distances.

When a telephone signal was transmitted over long distances through wires or cables, it was weakened or degraded due to factors like resistance, capacitance, and other electrical properties of the transmission medium. A repeater was placed at intervals along the transmission path to amplify the weak signal.

In addition to amplification, a repeater also regenerated the signal. That is, it cleaned up any noise or distortion that may have been introduced during transmission. The regenerated signal was then retransmitted, ensuring it remained clear and usable.

Telephone repeaters relied on electronic amplification techniques, particularly around the 1920s. While telephone repeaters were effective at amplifying signals, they had a significant drawback—they introduced distortion to the amplified signals. This distortion could affect the clarity and quality of the telephone conversations. Also, with the connection of many repeaters in sequence, the distortion would accumulate and degrade the signal to an unacceptable level. Another challenge was the expansion of telephone traffic on the network. As more and more people used telephones and made long-distance calls, the existing telephone lines became congested. Network operators needed to increase the lines' bandwidth to accommodate the increasing volume of calls. However, doing so came with a trade-off: higher transmission losses, which would diminish the quality of telephone communication.

Harold Stephen Black began working on this problem in the early 1920s, driven by the need to address the distortion issues associated with telephone repeaters. He recognised that traditional high-gain amplifiers, while effective at boosting signals, also introduced undesirable distortion due to noise and component drift.

On August 2, 1927, which marked a pivotal moment in the history of control engineering, Black ingeniously sketched a circuit for a negative feedback amplifier. This breakthrough concept involved feeding back a portion of the amplifier's output signal into its input. By doing so, he discovered that it was possible to reduce the distortion while maintaining amplification significantly. The invention was a *flash of insight* while Black was riding the Lackawanna Ferry across the Hudson River on his way to work. He recalled, "I felt an urge to write but had nothing to write on, so picked up my morning paper. By sheer coincidence one page was blank. Here was a perfect set-up, lots of room and fully dated. With this to implement my job, I started the first written record pertaining to the stabilised negative feedback amplifier. Years of study and many failures preceded this sudden conception of stabilised feedback. Despite immediate recognition of its importance, years of additional work were required before it found substantial commercial use" [39].

Following this discovery, extensive development efforts ensued. Full-scale practical trials of the negative feedback amplifier were conducted in 1930, marking a critical milestone in the evolution of telecommunications technology. By 1931, AT&T[10] had begun implementing this innovative amplifier within its communication systems, significantly improving signal quality and clarity.

Remarkably, information about the negative feedback amplifier remained undisclosed to the wider public until 1934. During developing this practical amplifier and the quest to understand its behaviour, Black received invaluable assistance from the brilliant engineer Harry Nyquist. Nyquist's seminal paper, Regeneration Theory, published in 1932, laid the foundations for what later became known as the *Nyquist analysis* [40]. This analysis became instrumental in telecommunications and control engineering, providing a powerful tool for the analysis and design of control systems. It also led to a methodology that did not require differential equations derivation and manipulations and for which experimental data—the measured frequency response—was sufficient to infer the system's stability and could estimate the stability degree.

Clesson E. Mason, working at the Foxboro Company[11] during the same period as Black, developed a pneumatic negative feedback amplifier. This innovation was applied to industrial control systems. Foxboro had previously invented the flapper-nozzle amplifier in 1914, but early versions of this technology were highly non-linear, effectively behaving like on–off switches. By the 1920s, attempts to modify it had only marginally improved its linearity, achieving about 7% of its full operational range.

In 1928, Mason began experimenting with feedback mechanisms to improve the linearity of the valve operation in these pneumatic amplifiers. By 1930, he had successfully developed a feedback circuit that linearised the valve's behaviour. This circuit allowed for introducing integral (or reset) action into the system's behaviour.

---

[10] AT&T, or American Telephone and Telegraph Company, is a historically significant telecommunications corporation in the United States. It has played a pivotal role in the development of telecommunications technology and services. Founded in 1885, AT&T has a rich history of innovation in the telecommunications industry. It was originally established as a subsidiary of the Bell Telephone Company, one of the companies formed by Alexander Graham Bell, who is credited with inventing the telephone.

[11] Foxboro was a prominent American industrial automation and control systems company. Founded in 1908, it played a significant role in the development and manufacturing of instrumentation and control devices for various industrial processes. The company made pioneering contributions to the field of industrial control.

As a result, in 1931, the Foxboro Company introduced the Stabilog pneumatic controller to the market. This controller incorporated linear amplification and integral action based on the negative feedback principle.

Initially, there was some resistance to adopting this new technology due to concerns about cost and a lack of understanding of how the controller operated. To address this, Foxboro produced a bulletin in 1932 explaining the principles of the system in simple terms and highlighting how it differed from *narrow-band*[12] controllers with limited linear ranges.

Simultaneously, at the Massachusetts Institute of Technology (MIT), Vanevar Bush oversaw extensive work on analogue calculating machines. This work led to the development of the differential analyser, which allowed for the simulation of dynamic systems and the numerical solution of differential equations. Additionally, Harold Locke Hazen and his MIT students worked on designing high-performance servomechanisms during this period. Hazen's theoretical studies, published in 1934, marked a significant milestone in control systems [41]. His work laid the groundwork for the next generation of control system specialists, contributing to the growing body of knowledge in control engineering and analogue computing.

2.6.4 The Advent of PID Controllers

The PID controllers are fundamental devices in control engineering used to regulate various processes and systems. They have a distinctive and prominent place in control engineering and are among the most widely used controllers up to the present day and have evolved and become a cornerstone of control theory and engineering practice. Many presentations, books, and technical papers deal with the history of PID controllers. However, [37, 42, 43] are the main sources used in this section.

On–off controllers were the dominant control strategy in the early twentieth century. In 1907, C.J. Tagliabue Co.[13] installed the first pneumatic automatic on–off temperature controller, and by 1910, both electrical and pneumatic on–off controllers had been further developed. The first on–off controller with a calibrated set-point scale was presented by Foxboro in 1912, and later, in 1927, Foxboro developed the pneumatic narrow-band proportional controller.

After the end of the First World War, there was a rapid increase in the utilisation of industrial instruments in the United States, leading to the development of practical PID pneumatic controllers. According to historical documents and professional assessments found in [42], it is claimed that between 1925 and 1935, more than

[12] Narrow-band controllers, also known as proportional controllers, are a type of control system used in process control and automation.

[13] C.J. Tagliabue Co. was a company that made significant contributions to the development of control systems, particularly in the field of temperature control, during the early part of the twentieth century. Founded in 1909 by Charles J. Tagliabue in New York City, the company initially focused on manufacturing thermostats and other temperature control devices for residential and commercial buildings.

75,000 automatic controllers were sold in the United States. In 1935, these controllers constituted 32% of the total sales for American instrument manufacturers. However, it is important to note that most of these controllers were simple on–off devices. Nonetheless, there was a growing recognition that, for many applications, achieving accurate control was not possible through simple on–off action.

The year 1920 marked a significant milestone in the history of control engineering with the granting of the first patent for a pneumatic Proportional-Integral (PI) controller to Morris E. Leeds, who was an American electrical engineer known for his many inventions in the field of electrical measuring devices and control. This particular achievement was noteworthy since it represents an early advancement in the field of control technology, specifically in using pneumatic systems for control purposes. Pneumatic control systems were prevalent during this era, and developing a PI controller added a new level of sophistication to industrial control processes.

The introduction of the pneumatic PI controller had practical implications for various industries. It allowed for more precise and stable control of processes, making it easier to maintain desired conditions in industrial applications such as temperature control, pressure regulation, and flow control.

Leeds' patent contributed to the development of control theory and the understanding of feedback control systems. The PI controller became one of the essential building blocks of modern control theory and practice.

PI control became known and widely used for its ability to reduce steady-state errors, but in certain cases, it led to oscillations, especially in systems with significant lag or delay. Confronted with this problem, Ralph Clarridge of the Taylor Instrument Companies[14] recalled an experiment where he restricted the feedback line of the proportional response controller. This adjustment effectively added some level of derivative control to the system. The experiment observed a significant *kick* or rapid change in the system's response when the setpoint was suddenly changed. This observation suggested that the controller was somehow *anticipating* the change in the error signal.

The term anticipating was used for the *derivative control action.* When the error signal changed rapidly, derivative control provided a corrective action that dampened the response and prevented overshooting or oscillations. Derivative control can be particularly effective in cases where there are sudden changes or disturbances in the system, helping to stabilise the response and reduce oscillations.

In 1935, Clarridge successfully implemented this restriction on the cellulose plant controller. After four years of its application in various plants, the Taylor engineers named this effect *pre-act.* The Taylor Instrument Companies introduced a completely redesigned version of its pneumatic controller and added pre-act or derivative control, as they labeled in 1939.

[14] In 1851, George Taylor started Taylor Instruments in Rochester New York. Taylor Instruments company specialised in manufacturing and distributing a wide range of measurement and control instruments for various industries. Today, Taylor Instruments offers a diverse product portfolio that includes pressure and temperature sensors, flow meters, level measurement instruments, and process control devices.

The same year, the Foxboro Instrument Company incorporated *Hyper-reset* alongside the proportional and reset control actions in their *Stabilog* pneumatic controller. Both pre-act and Hyper-reset actions introduced control mechanisms proportional to the error signal's derivative. Consequently, both controllers offered PID control.

Prior to the 1930s, control systems engineers and researchers were primarily engaged in the practical application of controllers across various plants and industries. The primary source of inspiration and guidance for control system design and development stemmed from the observations of human operators in real-world settings. The control designers tried mimicking the human operators' responses to closed-loop performance challenges, adopting purely heuristic approaches. This could indeed be regarded as an *intelligent* approach from a contemporary perspective.

Applications of expert systems in PID tuning are notable examples of the end of the last century [44]. With the emergence of expert system technology, capturing a control engineer's knowledge of the manual tuning of the PID controllers became feasible. Building an effective expert tuner for PI controllers was a research line followed by many control engineers.

Some control scientists and engineers noted the limitations of control systems at that time. In a 1934 paper [45], this situation is depicted as follows: "The science of the automatic regulation of temperature is currently in the anomalous position of having constructed an extensive practical structure on minimal theoretical foundations." It is also stated in a 1935 paper [46] that "In spite of the widespread use of automatic regulators and governors in all branches of industry, little has been published relative to the fundamental principles of the subject. In the absence of data necessary for rational solutions, the majority of automatic-control problems are solved at present by purely empirical methods." It is then claimed that twelve fundamental equations covering the characteristics of automatic regulators are presented, where six of these equations completely cover all the commercial regulators of that time.

PID controller tuning played a vital role in ensuring the effectiveness of control systems. With a significant interest in applying PID controllers across various industrial control challenges, tuning these controllers became necessary to harness their capabilities thoroughly. It became evident that different processes and systems exhibited unique characteristics, such as inertia, delay, and response time. Achieving precise and stable control required fine-tuning PID controllers to align with the specific dynamics of the systems under their control.

Tuning methods often relied on empirical rules and guidelines based on practical experience and developed heuristics. These rules served as initial starting points for tuning. A significant early contribution was the Ziegler-Nichols method, introduced in the 1940s [47]. This method provided systematic procedures for tuning PID controllers based on process reaction to step changes in control output. It established a framework for selecting proportional, integral, and derivative parameters.

Interestingly, Aidan O'Dwyer, in his work *PID Control: The Early Years*, notes that the first-ever tuning rule for controllers was actually derived eight years before the publication of the Ziegler and Nichols paper. This pioneering tuning rule was documented in a 1934 ICI[15] technical report authored by Callander, an engineer at ICI. The report was archived in the Central Filing Department of ICI (Alkali) Ltd., Northwich, and was fortunately acquired by O'Dwyer in 2004. For more details, refer to [48].

2.7 Conclusions

Studying the history of control systems design is invaluable as it provides insights into the evolution of control theories and design concepts, highlighting the intellectual journey that has shaped our current understanding. This chapter briefly outlines the key breakthroughs in engineering control systems from antiquity to the pre-industrial and pre-classical control era. It is evident that during these eras, control systems heavily relied on skilled craftsmanship, and selected achievements were presented. The integration of mathematical principles into control theory, particularly in the calculus of variations and mathematical stability analysis, is explored. Although the theory of calculus of variations did not formally enter control systems analysis and design, it laid the foundations for optimality in control. Furthermore, the history of control theories reveals paradigm shifts in control science, where fundamental theories and practices gave way to new concepts, fundamentally altering system analysis and design approaches. Examining these shifts provides valuable insights into the evolution of control science and the acceptance of innovative ideas.

References

1. H.S. Black, Stabilized feedback amplifiers. Bell Syst. Tech. J. **13**(1), 1–18 (1934)
2. J.G. Landels, Water-clocks and time measurement in classical antiquity. Endeavour **3**(1), 32–37 (1979)
3. A.M. Lepschy, G.A. Mian, U. Viaro, Feedback control in ancient water and mechanical clocks. IEEE Trans. Educ. **35**(1), 3–10 (1992)
4. https://mathshistory.st-andrews.ac.uk/Biographies/Heron/. Accessed 24 Dec 2024
5. C.H. Houpis, S.N. Sheldon, J.J. D'Azzo, *Linear Control System Analysis and Design: Revised and Expanded* (CRC Press, 2003)
6. https://historycollection.com/ancient-technology-centuries-time/. Accessed 24 Dec 2024
7. H.-S. Yan, *South-pointing Chariots,* in *Reconstruction Designs of Lost Ancient Chinese Machinery* (History Of Mechanism And Machine Science) (Springer Science & Business Media, 2007)
8. F.L. Lewis, D. Vrabie, V.L. Syrmos, *Optimal Control* (John Wiley & Sons, 2012)
9. https://muslimheritage.com/a-review-of-early-muslim-control-engineering/. Accessed 24 Dec 2024
10. https://www.1001inventions.com/top7clocks/. Accessed 24 Dec 2024

[15] ICI, or Imperial Chemical Industries, was a British chemical company that existed for much of the twentieth century. It was one of the world's largest chemical manufacturers and played a significant role in the chemical industry and various other sectors.

11. G. Nadarajan, Islamic Automation: Al-Jazari's Book of Knowledge of Ingenious Mechanical Devices, Muslim Heritage. http://www.muslimheritage.com/article/islamic-automation-al. Accessed 30 Dec 2024
12. M. Shatzmiller, Ahmad Y. al-Hassan et Donald R. Hill, Islamic technology. An illustrated history. Paris, Cambridge University Press and Unesco, 1986. Bull. Critique des Ann. Islamologiques, **6**(1), 206–208 (1989)
13. D.R. Hill, *The Book Kitāb al-Hiyal"of Banū Mūsā bin Shākir: Interpreted in Sense of Modern System and Control Engineering. Atilla Bir, Ekmeleddin Ihsanoglu* (1992)
14. L.L. Schmidt, Islamic automata in the absence of wonder, Massachusetts Institute of Technology, 2010
15. A. Bir, M. Kacar, Pioneers of automatic control systems Essays Honour Ekmeleddin İhsanoğlu Soc. Cult. Sci. Collect. Art. (2006)
16. P. Hill, *The Book of Knowledge of Ingenious Mechanical Devices: (Kitāb fī ma'rifat al-ḥiyal al-handasiyya)* (Springer Science & Business Media, 2012)
17. https://whc.unesco.org/en/tentativelists/6192/. Accessed 24 Dec 2024
18. https://www.britannica.com/technology/windmill. Accessed 24 Dec 2024
19. https://www.britannica.com/biography/Edmund-Lee. Accessed 24 Dec 2024
20. O. Mayr, The origins of feedback control. Sci. Am. **223**(4), 110–119 (1970)
21. E. Gallo, Jean Simon Bonnemain (1743–1830) and the Origins of Hot Water Central Heating, in *Proceedings of the Second International Congress on Construction History* (Construction History Society, 2006), pp. 1043–1060
22. E. Jury, On the history and progress of sampled-data systems. IEEE Control Syst. Mag. **7**(1), 16–21 (1987)
23. H.J. Sussmann, J.C. Willems, 300 years of optimal control: From the brachystochrone to the maximum principle. IEEE Control Syst. Mag. **17**(3), 32–44 (1997)
24. N. Rouche, P. Habets, M. Laloy, *Stability Theory by Liapunov's Direct Method* (Springer, 1977)
25. J.C. Maxwell, On governors. Proc. Roy. Soc. Lond. **16**, 270–283 (1868)
26. N. Wiener, *Cybernetics or Control and Communication in the Animal and the Machine* (MIT Press, 1961)
27. C.-G. Kang, Origin of stability analysis: on governors by JC Maxwell [Historical Perspectives]. IEEE Control Syst. Mag. **36**(5), 77–88 (2016)
28. E.I. Jury, Remembering four stability theory pioneers of the nineteenth century. IEEE Trans. Autom. Control **41**(9), 1242 (1996)
29. A.N. Michel, Stability: the common thread in the evolution of feedback control. IEEE Control Syst. Mag. **16**(3), 50–60 (1996)
30. C. Bissell, Russian and Soviet contributions to the development of control engineering: a celebration of the Lyapunov centenary. Trans. Inst. Meas. Control **14**(4), 170–178 (1992)
31. A. M. Lyapunov, *General problem of the stability of motion* (no. 3). CRC Press, 1992.
32. P.C. Parks, AM Lyapunov's stability theory—100 years on. IMA J. Math. Control Inf. **9**(4), 275–303 (1992)
33. R.E. Kalman, J.E. Bertram, Control system analysis and design via the "second method" of Lyapunov: I—Continuous-time systems. J. Basic Eng. (1960)
34. S. Lefschetz, La Salle, J.P, *Stability of Lyapunov's Direct Method with Applications* (Academic Press, New York, 1961)
35. A. Fuller, A history of control engineering 1800–1930. Electron. Power **25**(9), 651 (1979)
36. S. Bennett, A brief history of automatic control. IEEE Control Syst. Mag. **16**(3), 17–25 (1996)
37. S. Bennett, *A History of Control Engineering, 1930–1955* (no. 47) (IET, 1993).
38. N. Minorsky, Directional stability of automatically steered bodies. J. Am. Soc. Naval Eng. **34**(2), 280–309 (1922)
39. H. Black, Invention in engineering. Electr. Eng. **77**(8), 722–723 (1958)
40. H. Nyquist, Regeneration theory. Bell Syst. Tech. J. **11**(1), 126–147 (1932)
41. H.L. Hazen, Theory of servo-mechanisms. J. Franklin Inst. **218**(3), 279–331 (1934)
42. S. Bennett, Development of the PID controller. IEEE Control Syst. Mag. **13**(6), 58–62 (1993)
43. S. Bennett, The past of PID controllers. Ann. Rev. Control **25**, 43–53 (2001)

44. J. Lieslehto, J. Tanttu, H. Koivo, An expert system for tuning PID controllers, in *Computer Aided Design in Control Systems*. Elsevier **1989**, 341–346 (1988)
45. A. Ivanoff, Theoretical foundations of the automatic regulation of temperature. J. Inst. Fuel **7**(33), 117–130 (1934)
46. S.D. Mitereff, Principles underlying the rational solution of automatic-control problems. Trans. Am. Soc. Mech. Eng. **57**(4), 159–163 (1935)
47. J.G. Ziegler, N.B. Nichols, Optimum settings for automatic controllers. Trans. Am. Soc. Mech. Eng. **64**(8), 759–765 (1942)
48. A. O'Dwyer, PID control: the early years, Technological University Dublin, ARROW@TU Dublin (2005)

Chapter 3
A Concise History of Control Systems: From Classical to Modern Control Era

Much remains to be done. But the Promised Land has been sighted from several directions.
—Rudolf Kalman

3.1 Introduction

Significant theoretical and practical advancements in automatic control emerged between the late 1930s and early 1960s. This chapter outlines key historical milestones of *classical control theory,* the set of control techniques and methods developed predating *modern control theory*, which includes state-space and optimal control methods.

Frequency domain analysis techniques were widely used in classical control theory. Engineers used these tools to analyse and design control systems to attain closed-loop systems with specified frequency response characteristics. Also, the s-domain root locus method, a graphical technique for analysing control systems' stability and transient response, was developed and extensively used during this era. It allowed engineers to design controllers for desired closed-loop performance.

The concept of transfer functions, which describes the relationship between the input and output of a linear time-invariant system (LTI), became a fundamental part of classical control theory. Engineers used transfer functions to model and analyse LTI dynamical systems. Classical control theory became a core subject in engineering education during this period.

Continuing the achievements of this flourishing period, one of the most fundamental developments was the introduction of state-space representation or the emergence of the modern control period. This mathematical framework allowed engineers and scientists to describe dynamic systems more abstractly and generally. Also, by utilising state-space concepts and tools, mathematicians and engineers defined and made substantial contributions to *optimal control theory* during this period. *Optimality* addressed the important problem of finding control inputs that optimised a certain criterion while considering system dynamics and constraints.

A. Khaki-Sedigh, *Control Systems: A Historical and Philosophical Perspective*,
https://doi.org/10.1007/978-981-96-7426-8_3

Researchers focused on developing rigorous stability analysis techniques to ensure control systems operate safely and predictably. With the advent of digital computers, there was a shift toward digital control systems, offering improved precision and flexibility in control algorithms.

On the other hand, from an application perspective, the demands of World War II accelerated the development and application of automatic control systems, particularly in aviation and weaponry. Automatic control was crucial in stabilising and guiding aircraft, missiles, and anti-aircraft systems.

In the post-war era, industries embraced automatic control to improve manufacturing processes, increasing efficiency and product quality. Feedback control systems were implemented in factories to maintain consistent production.

The 1950s and 1960s marked the Space Race between the United States and the Soviet Union. Developing guidance and control systems for rockets and spacecraft became a critical focus, pushing the boundaries of automatic control.

The growth of emerging technologies, such as telecommunications, nuclear power, and aerospace, required sophisticated control systems to ensure safety and reliability.

Chapter 2 discussed the history of control system design from antiquity through the pre-industrial period and into the pre-classical control era. This chapter explores the period during which mathematical principles of control systems were established within the context of classical control techniques. We will then progress to the era that represents the advancements made in the twentieth century, marked by the introduction of state-space mathematical control system theory at its core. Rudolf Kalman, often regarded as the father of modern control theory, referred to the achievements of this era as the *Promised Land* [1].

3.2 Classical Control Period

3.2.1 The Foundations of Classical Control Systems

In the late 1930s, two clearly defined and well-established approaches existed for examining the behaviour of feedback systems:

1. The time-response method, which utilised ordinary differential equations alongside their corresponding algebraic properties.
2. The frequency–response approach, which covered Nyquist and Bode plots, transfer functions, and associated tools. This technique was primarily used to analyse feedback amplifiers.

The frequency–response approach used *transfer functions* to describe the input–output behaviour of a system, representing systems as blocks where an input signal entered a block and produced an output signal. These blocks were arranged into *block diagrams* to depict closed-loop systems, serving as a universal language for engineers from different disciplines to communicate and share their control systems analysis and design experiences across various applications.

A detailed account of the history of frequency–response techniques and their implications in automatic control systems analysis and design can be found in [2, 3].

The classical Nyquist paper [4] represents a truly historic event in the evolution of frequency–response methods. Nyquist's research focused on the stability issue of feedback amplifiers used in long-distance telephony. This stability problem stemmed from the feedback amplifiers designed by Black, which were discussed in Sect. 2.6.3. Remarkably, Black coined the term *singing* to describe the instability phenomenon observed in his feedback amplifiers. Telephone engineers handling audio signals noted that amplifiers tended to exhibit singing behaviour when the loop gain was increased, and, in some cases, singing unexpectedly occurred when the loop gain was decreased.[1] Additionally, the feedback amplifier system had a very high order, comprising approximately 50 independent energy storage elements, including inductors and capacitors. This made Maxwell's time-response differential equations-based stability analysis approach infeasible and non-applicable.

Nyquist's open-loop gain frequency–response solution to the feedback stability problem held immense practical value. By expressing the problem in terms of a directly measurable gain, this technique introduced an essential link with experimental measurements in practical control applications, thereby enhancing its relevance and applicability. Notably, Nyquist's stability criterion did not necessitate a mathematical model in the form of differential or characteristic equations. Furthermore, based on open-loop gain and frequency behaviour, it provided a clear measure for enhancing a system's feedback stability and stabilising it when dealing with unstable closed-loop systems.

As the Nyquist achievement could significantly impact long-distance telephony, Bell Laboratory engineers conducted thorough experimental investigations to examine the proposed Nyquist stability criterion rigorously. Fortunately, the experimental results confirmed Nyquist's theoretical stability assessments, leading to a substantial elevation of frequency–response techniques within the control engineering community.

Another remarkable point is that before the introduction of the Nyquist stability criterion, the notion of *conditional stability* was unknown to control engineers. The prevailing belief was that for a given system, there existed a single critical gain value that delineated stable and unstable behaviours. Black in [5] describes the phenomenon of conditional stability in negative feedback amplifiers and concludes that "Results of experiments, however, seemed to indicate something more was involved and these matters were described to Mr H. Nyquist, who developed a more general criterion for freedom from instability applicable to an amplifier having linear positive constants".

The above developments occurred in the telephone engineering domain. The transfer of these ideas to other fields became possible through the introduction of transfer functions in the analysis of general feedback systems by Harris at MIT in

---

[1] This differed from the case of governor hunting observed by Watt and studied by Maxwell.

1942 [6]. Harris's idea enabled a servomechanism or a control system in other engineering disciplines to be represented in block diagram terms and thus analysed using the powerful tools available to the feedback amplifier designers [2].

Another significant contributor to the frequency response analysis was Hendrik Bode at Bell Labs. He began working on equalisers in the mid-1930s. Engineers at Bell Labs were deeply engaged in efforts to expand the bandwidth of their communication systems while achieving desired frequency response characteristics. Their ideal goal or *dream* was to attain constant gain across a wide bandwidth, along with a sharp cut-off and minimal phase lag. In other words, regardless of the signal's frequency passing through the system, they wanted the amplification or attenuation to remain consistent. This was crucial to ensure that signals were not distorted or attenuated unevenly as they travelled through the system.

A wide bandwidth implied that the system could effectively transmit signals at various frequencies. This is essential for accommodating different signals and data rates in communication systems. The engineers also aimed for a sharp cut-off in their systems so that the system would significantly attenuate or block signals beyond a certain frequency. All of this should have been achieved with minimum phase lag to ensure that the output signal closely follows the input signal in terms of timing.

However, despite the diligent efforts of Bell Labs engineers, they discovered that achieving the desired gain characteristic resulted in excessive phase lag. In 1940, Bode demonstrated that there is no universally applicable relationship between attenuation and phase shift for a physical structure. Instead, he revealed a connection between a specific attenuation characteristic and the minimum phase shift associated with it. It was Bode who demonstrated that the *dream* of telephone engineers to construct circuits with rapid cut-off and minimal phase shift was, as expressed by Bissell, indeed a *dream* [7].

Notably, Bode introduced the critical point $(-1, 0)$, which is still in use today, as opposed to Nyquist's choice of the point $(+1, 0)$. He also introduced the concepts of gain and phase margins and the gain-bandwidth limitation. Bode's work included the famous straight-line approximations for frequency response curves of linear systems plotted on semi-log axes. Comprehensive details of Bode's contributions can be found in his 1945 book [8].

During the same period, Harold L. Hazen and Gordon S. Brown at MIT pursued time-domain techniques based on the operator approach. They developed and employed block diagrams for system analysis, pioneering the use of the mechanical differential analyser to simulate control systems for the first time. In 1934, Hazen published a seminal paper on the theory and design of servomechanisms, providing simple descriptions of servo operation and a methodology for design [9].

A differential analyser was an analogue computer used for solving differential equations, particularly those involving mechanical systems, and was later used for simulating control systems. It was a pioneering device in the history of computing and played a crucial role in engineering and scientific calculations before the advent of digital computers. A (mechanical) differential analyser includes a physical mechanism consisting of a complex arrangement of gears, shafts, and linkages. These mechanical components were used to mimic the mathematical operations of differentiation and integration. The core of the differential analyser was a series of interconnected rotating shafts, with each shaft representing a variable in the differential equation being solved. The machine can perform integration and differentiation operations by connecting the shafts in various ways and using gears of different sizes.

In 1936, a differential analyser developed as a joint research project between MIT and the University of Manchester in the UK was used by Callender[2] of ICI for the first time to simulate an industrial control system.

3.2.2 The Effects of World War II on Automatic Control Systems

World War II tremendously affected the advancement of automatic control systems. Before the war, as discussed in Sect. 3.2.1, design techniques and control devices had been developed as part of the work on electronic amplifiers, electrical filters, differential analysers, and process controllers. The demands of war focused attention on other challenging control problems. This issue is discussed in many papers. For a comprehensive treatment of the subject, refer to the following sources [10–14], and Chap. 1 of [15].

These war-challenging problems can be categorised as follows. The development of advanced weaponry, including guided missiles and torpedoes, required precise control systems to navigate and target. The war accelerated the research and development of guidance and control systems, leading to innovations in guidance algorithms and control hardware. World War II saw the rapid advancement of military aviation. Control systems were critical in improving aircraft stability, manoeuvrability, and targeting accuracy. Automatic pilots and flight control systems became essential for long-range bombers and fighter aircraft.

The need for effective anti-aircraft defence systems led to the development of radar-guided tracking and targeting systems. These systems relied on automatic control to track and intercept enemy aircraft, significantly advancing radar technology and control algorithms.

Control systems were instrumental in artillery fire control, enabling more accurate targeting and increased firing rates. Sophisticated fire control systems were developed to improve artillery's effectiveness in offensive and defensive operations.

In defensive artillery fire control for anti-aircraft systems used to intercept attacking bombers, it was imperative to determine their position accurately. This task was accomplished through optical rangefinders and tracking telescopes, which provided information about the target's range, bearing, and elevation.

As the war progressed, radar technology gradually assumed these functions, initially focusing on rangefinding and eventually incorporating active tracking capabilities.

[2] Refer to Sect. 2.6.4 for Callender.

Subsequently, it became essential to calculate the target's anticipated future position, essentially computing a *lead*, based on the aircraft's direction and velocity. The gun director was responsible for predicting the future location of the target by factoring in its speed and direction. It then calculated output values for azimuth and elevation, which were used for aiming the guns, and a fuze setting that determined the time delay after firing for the shell to explode.

The data generated by the gun director was transmitted to the gun itself to control its movement precisely. The gun could be automatically adjusted using hydraulic or electric power controls or manually operated based on *follow-the-pointer* indicators, which allowed gunners to align the gun accurately.

The central computer or *gun director*, had the critical role of integrating all available data while accounting for variables such as wind, terrain, and specific ballistics characteristics that depended on the particular gun and shell being used. This complex operation demanded the coordination of up to 14 personnel responsible for observing and tracking tasks.

Designing an effective servomechanism to control the gun's position presented significant challenges. It became evident in 1941 that the labour-intensive process of manually relaying information from radar devices to gun controllers was inadequate for countering the threat posed by fast-moving aircraft. Consequently, there was a compelling need to develop a system in which an automatic tracking radar was directly linked to the gun director, which, in turn, was linked to the gun position controller [10]. A schematic of the human in the loop-guided gun system is depicted in Fig. 3.1.

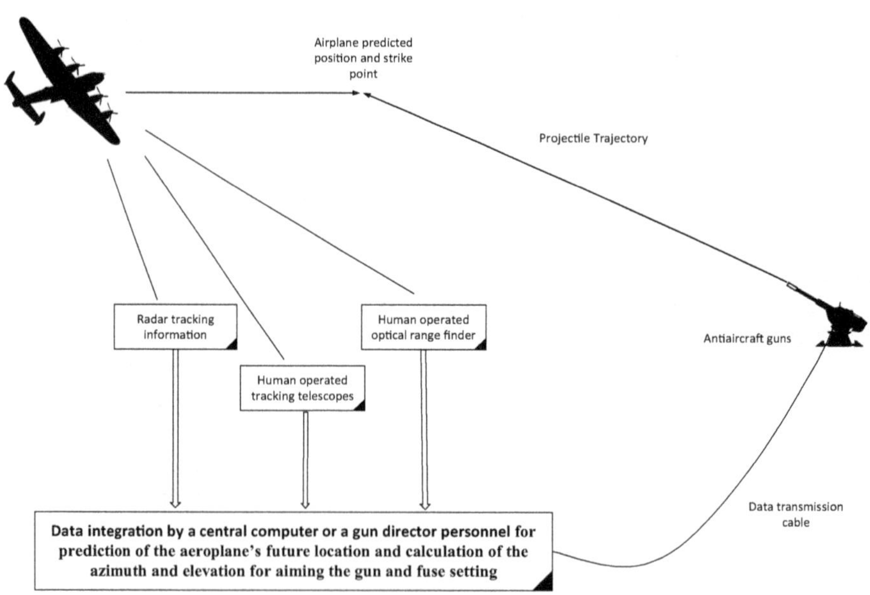

Fig. 3.1 An artillery fire control system with human operators in the feedback loop

This may well be the initial encounter of engineers with systems that amalgamated mechanical, electrical, and electronic engineering disciplines into a *unified* closed-loop system. One notable outcome of this convergence was the emergence of a *systems engineering* perspective. Hazen noted [10] that a fire-control system was more than the sum of component parts and saw it as an integrated whole with interrelated functioning of all its parts. He also appreciated the systematic approach by arguing that "one is safe in considering the parts separately only if one always keeps in mind their relation to the whole."

The NDRC, or National Defense Research Committee, was the United States government agency established during World War II. It was created on June 27, 1940, by President Franklin D. Roosevelt to coordinate scientific research for military purposes. The NDRC played a crucial role in harnessing the expertise of scientists, engineers, and researchers to support the war effort. It was responsible for conducting and overseeing various research projects related to defence, including automatic control systems in defence systems. Many prominent scientists and engineers were involved with the NDRC, including Vannevar Bush, who was the committee's chairman. The work of the NDRC had a profound impact on the outcome of the war and laid the groundwork for post-war scientific and technological advancements.

A key project proposed and led by the NDRC during the war was the development of an artillery fire control system, which involved the core challenge of predicting the movement of approaching aeroplanes. In fact, the NDRC simultaneously signed contracts with Bell Labs and Norbert Wiener at MIT to address the prediction problem.

Wiener proposed a groundbreaking approach to mathematically address the challenge of predicting the future position of aircraft, marking the first mathematically defined prediction problem in this context. His proposal was built upon his work in the 1920s on *generalised harmonic analysis.*[3]

Collaborating with John Bigelow, they successfully developed an *electronic prediction system*. Bigelow was an American engineer and scientist who played a significant role in the development of radar technology during World War II, particularly in the NDRC project. While Norbert Wiener is often more prominently recognised for his contributions to the mathematical foundations of prediction systems, John Bigelow played a vital role in translating these mathematical ideas into practical electronic systems for military use.

However, Wiener was disappointed that, in the end, his system achieved only a marginal improvement (less than 10%) over the system developed empirically by the Bell Telephone Laboratories.

The work resulted in Wiener producing a report titled *The Extrapolation, Interpolation, and Smoothing of Stationary Time Series with Engineering Applications,*

[3] In the field of generalised harmonic analysis, Wiener extended the principles of harmonic analysis, which involves decomposing functions into their constituent sinusoidal or harmonic components, often using techniques such as Fourier analysis. What sets generalised harmonic analysis apart is its ability to handle functions that may not be inherently periodic or amenable to conventional Fourier analysis methods. This discipline encompasses a broader range of functions and employs diverse mathematical tools for their analysis and representation.

published as an OSRD Report in 1942. The OSRD, or Office of Scientific Research and Development, was a United States government agency established during World War II. Its primary purpose was coordinating and supporting scientific research and development efforts related to military technology and wartime needs. This report was called the *Yellow Peril* due to its distinctive yellow cover and its formidable mathematical challenges. The report was later published in the open literature in 1949.

Willems has an interesting point on this in [16], where he notes that Wiener's World War II-era attempts to apply stochastic filtering to aircraft trajectory prediction reveal fundamental limitations in modeling tactical evasion probabilistically. While the approach assumed pilot maneuvers could be represented as a stationary stochastic process, this framework failed to account for the fact that deliberate evasion follows tactical logic rather than probability distributions, and pilots adapt maneuvers based on observed threats. This conceptual mismatch between mathematical idealisation (stationary noise processes) and combat reality (goal-directed evasion) likely explains the approach's operational shortcomings. He then concludes that "the widespread use of probability as a panacea for uncertainty, as a matter-of-fact and universal model that needs no justification nor a discussion regarding its interpretation, is for me a constant source of *Unbehagen.*[4]"

World War II also saw the birth of electronic warfare. Control systems were used to jam enemy communication and radar systems, marking the beginning of electronic countermeasures. They were vital in navigating submarines and underwater vehicles and helped submarines maintain depth, course, and position, enhancing their stealth and effectiveness.

Analogue computers were developed and used extensively during the war to solve complex control problems. These early computers facilitated calculations for trajectory prediction, target tracking, and other control-related tasks.

The wartime environment necessitated collaboration between engineers, mathematicians, physicists, and military experts. This cross-disciplinary cooperation led to rapid advances in control theory, including the development of feedback control principles.

Many control systems technologies and knowledge developed during the war transitioned into civilian applications. This included adapting military control systems for use in commercial aviation, manufacturing, telecommunications, and other industries.

[4] *Unbehagen* is a German word that translates to *discomfort* or *unease* in English. Thus, "a constant source of Unbehagen" implies that the use of probability as a universal model without justification or discussion about its interpretation is a continual cause of discomfort or unease.

3.2.3 The Birth of Cybernetics

Wiener's work during the war proved immensely influential in his later work on *cybernetics*. Furthermore, the problem of *smoothing* mentioned in the title of his report was connected closely to that of *prediction*. Most gun directors predicted the target's future position by linearly extrapolating velocity. But when attacking pilots took evasive action, the target could have been lost. Wiener took a statistical approach to this problem, building his *prediction machine*. However, as mentioned earlier, Wiener's work was too theoretical, with only a slight advantage to be of any use for the war. Bell Labs' engineers recognised the value of Wiener's work but rejected its statistical approach in favour of more standard solutions. Still, Wiener's fire control research shaped his later work on cybernetics.

In 1948, Wiener published his book *Cybernetics* [17], drawing heavily from his experiences and insights gained through his work with the NDRC. Wiener's vision of *human–machine interactions* was shaped by his involvement in prediction systems and his fascination with biology. His contributions left a lasting impact on the *cybernetic* research community, which had been exploring earlier concepts of human–machine interactions. In [15], Wiener's work is portrayed as the efforts of a tweedy scholar with thick, horn-rimmed glasses who uncovered the magic of feedback loops, self-stabilising systems, and machines capable of autonomously adapting their behaviour and learning. Automata gained a purpose and, at least theoretically, the ability to self-reproduce. The machine suddenly appeared lifelike.

The first industrial revolution was characterised by replacing human muscles with machines, such as steam engines and assembly lines. Wiener in the introduction of [17] writes "the first industrial revolution, the revolution of the *dark satanic mills*, was the devaluation of the human arm by the competition of machinery." In contrast, developing automation and control systems can be seen as a foundation for the modern technological landscape. Wiener's work on cybernetics explored the potential applications of automation and control systems, laying the groundwork for the widespread use of artificial intelligence, robotics, and computer networks. In referring to the *Second Industrial Revolution*, Wiener argues that the modern industrial revolution is also likely to diminish the value of the human brain, particularly in handling simpler and more routine decisions. He continues to note that automation would require more highly skilled scientists and administrators. Lower-skilled workers will face increased competition in the job market, leading to a higher risk of unemployment.

In a final note on Wiener's contribution to control theory, he recognised that the autocorrelation function was a crucial link between the time and frequency–response descriptions of stochastic processes. Building on this insight, he conducted classic studies in random process analysis, exploring their relationships with communication and control theory using the generalised Fourier transform of this function.

Wiener's deep interest in the connections between control and communication problems and his recognition of similarities between such problems in engineering and physiology had a profound impact. His efforts played an important role in disseminating feedback-control concepts and frequency–response methods into the fields of stochastic system theory and physiology.

Within the cybernetics relam, exploring feedback mechanisms, machine control, and the role of human operators transformed our understanding of information and complex systems. Notably, several researchers from the NDRC, inspired by Wiener's pioneering work, went on to develop their own *cybernetic visions*, which, while aligned with Wiener's ideas, also bore distinct characteristics and contributions of their own.

For instance, Claude Shannon investigated fire control and cryptography while at Bell Labs. He ingeniously merged these two domains—machine control and messaging—in his 1948 paper titled *A Mathematical Theory of Communication* [18].

Also, Jay Wright Forrester, who worked on developing servo mechanisms for aircraft control during the war, was associated with the Servo Lab at MIT, which played a crucial role in advancing control systems technology during the war as part of the NDRC. Forrester was a prominent American engineer and systems scientist known for pioneering system dynamics work. He contributed significantly to studying complex systems, feedback loops, and their applications in various disciplines, including engineering, economics, and social sciences.

After the war, Forrester continued his work in control systems and began developing the field of *system dynamics*. He introduced the concept of feedback loops and dynamic modelling to analyse and simulate complex systems over time. His work in this area led to the creation of the first computer-based system dynamics model called the *Whirlwind Computer*.[5] Forrester's contributions to system dynamics and his innovative approach to understanding complex systems continue to influence researchers and practitioners across multiple disciplines. His cybernetics vision has impacted fields such as management, economics, environmental studies, and policy analysis. Forrester's dedication to advancing the understanding of dynamic systems and his development of system dynamics as a field of study have significantly affected our perception of how to analyse and address complex problems in various domains.

3.2.4 Moving Beyond Classical Control Systems

As discussed in Sect. 3.2.1, the predominant conceptual framework of dynamical systems during the early to mid-twentieth century was rooted in input/output and frequency-domain methodologies. In this specific era, transfer functions played a central role, serving as the primary method for characterising systems under control.

[5] The Whirlwind Computer was a pioneering digital computer developed at the MIT during the late 1940s and early 1950s. It was one of the earliest high-speed, real-time digital computers and is particularly notable for its role in advancing the field of real-time computing and air defence systems.

Symbolic calculus, introduced by Heaviside, provided a powerful tool to handle complex systems through a combination of series, parallel, and feedback interconnections. This method facilitated the conceptualisation of differential operators or delays as formal indeterminates, offering a flexible and intuitive approach.

By the early 1950s, classical control techniques were the dominant strategy for practical control problems. They were the routinely used powerful tools for analysing and designing feedback mechanisms and automatic control systems. The *input–output* or *external description* of dynamical systems with transfer functions and block diagram representations was widely employed by control engineers handling many different problems in various fields of engineering. Classical control techniques were instrumental in providing engineers with a common language and communication medium, enabling them to tackle complex real-world challenges collaboratively. In essence, these techniques transformed control engineering and promoted interdisciplinary collaboration, marking a key point in the history of control systems.

However, the mathematical formalism of Laplace transforms and complex functions presented challenges. As noted in [16], analysts had constrained Wiener's vision of cybernetics and feedback systems within the rigid framework of Laplace transforms, employing complex functions with domains of convergence and other cumbersome mathematical obstacles that were largely irrelevant to the core principles of adaptive, purposeful machines.

The dominant framework for LTI systems faced significant limitations, stalling progress in control theory. Without a rigorous mathematical treatment of polynomials and rational functions, researchers struggled to advance into more complex areas, particularly linear multivariable systems. From various perspectives—modelling, physical understanding, and mathematical formulation—control theory had a somewhat *narrow* and *shallow* scope. This limited perspective made the comprehensive exploration of dynamic systems difficult. Recognising the challenge, a call emerged for a more sophisticated approach that could transcend existing constraints.

Despite the notable success of Nyquist and Bode diagrams in representing physical systems based on plant response measurements and the usefulness of Evans' root-locus method as a powerful tool for designing high-order linear feedback systems, as well as advancements in managing stochastic disturbances, limitations within classical control theory were becoming apparent.

The key factors that necessitated and facilitated the move beyond classical control can be summarised as follows:

- The widespread availability of detailed mathematical models for controlled plants in the form of differential and algebraic equations.
- The increasing importance of optimising control systems for economic efficiency, which become a central concern for control engineers.
- Tackling complex problems associated with multivariable, large-scale systems, moving beyond the scope of feedback amplifiers and single-loop servomechanisms.

- The advent of stored-program digital computers and specialised computing systems provided reliable and widely accessible computing power, enabling control engineers to explore more sophisticated control strategies.
- As industries and technologies evolved, there was a growing need to integrate control systems with other aspects of engineering and automation, such as communication networks, sensors, and advanced instrumentation. This integration required more flexible and adaptable control strategies.
- With increasing environmental and safety regulations awareness, control systems were needed to meet stringent requirements, such as emissions control and safety interlocks.
- In a globalised market, industries faced stiff competition and needed more sophisticated control systems. This drove the adoption of advanced control strategies to optimise processes, reduce costs, and improve product quality.
- The development of advanced sensors and actuators allowed control engineers to measure and manipulate systems in ways that classical control methods could not fully exploit.

As a clear illustration of some of the previously mentioned points, Gordon Brown and Duncan Campbell[6] outlined their vision for the future of control applications in their 1949 work, as cited in [19]. They articulated a forward-looking vision for control applications. They emphasised that true progress in automation required integrated system design with coordinated design of plants, instruments, and control equipment. Their philosophy aimed to promote product quality, optimise plant coordination, clarify economic trade-offs in design, and ensure safety within industrial societies. For example, they noted that industries with large-scale production could achieve significant gains in efficiency and product consistency through advanced control methods. Other critical drivers included raw material conservation, energy efficiency, and mitigating environmental hazards like air and water pollution. These were classified as challenges where control engineering could address both humanitarian and economic imperatives.

3.3 State-Space-Based Mathematical Control Theory Period

The study of science can be broadly categorised based on the complexity of the systems under investigation. The natural sciences typically examine systems with limited components, where fundamental properties and interactions are well-defined. In contrast, disciplines such as cybernetics and systems engineering address

[6] Gordon Stanley Brown (August 30, 1907–August 23, 1996) was an Australian-born professor of electrical engineering known for his significant contributions to the field and Donald P. Campbell was his student. He spent much of his career at the Massachusetts Institute of Technology (MIT). Brown played a key role in the development of concepts related to automatic-feedback control systems and the numerical control of machine tools.

phenomena characterised by high system-component complexity. These systems, whether natural or artificial, comprise numerous interconnected elements that interact in intricate ways. The essential distinction between system categories lies in the scale and complexity of their *system-components*.

To handle these complexities, the concept of a *model* was introduced. In this context, a model is an abstract construction, a set of rules or representations that serves several crucial functions. It aids in the *analysis* of intricate problems by breaking them down into manageable components and guides *how to achieve* specific objectives within these complex systems.

For the analysis and control of natural or man-made complex systems, the obvious step forward would be the development of system thinking or *system theory*. System theory emerged to defy simplification through straightforward laws or principles. In essence, system theory is an indispensable discipline for handling the intricate complexities of systems marked by a substantial system component [20]. The model and system theory issues are discussed in more detail in Chaps. 5 and 6.

3.3.1 Poincaré's Impact on Dynamical Systems Study

The mathematical control theory that emerged in the late 1950s and early 1960s is based on time-domain analysis by describing dynamical systems in terms of differential and algebraic equations. Henri Poincaré a French mathematician and physicist made significant contributions to various areas of mathematics and science in the late 19th and early twentieth centuries. His work in the field of dynamical systems, in particular, had a profound and lasting impact.

Poincaré emphasised the importance of describing dynamical systems using first-order differential equations. These equations express how the *state* of a system changes over time based on its current *state*. By formulating theories in this framework, he provided a powerful mathematical foundation for modelling various dynamic phenomena.

Poincaré introduced the concept of considering the relevant set of dynamical system variables as the trajectory of a point in an n-dimensional space, often referred to as *phase space*. In this n-dimensional space, each dimension represents a different variable or *state variable* of the system. The trajectory of the point in this space represents the system's *evolution* over time. This geometric approach allowed for a visual representation of complex system behaviour.

Poincaré's studies led to the development of *chaos theory*, which deals with the behaviour of highly sensitive systems to initial conditions. Poincaré also introduced the concept of a *Poincaré map*, which involves projecting a system's trajectory onto a lower-dimensional cross-section of the phase space. This technique simplifies the analysis of complex systems and helps identify important features such as stable and unstable periodic orbits.

Poincaré's emphasis on using differential equations and phase space representation revolutionised the way the behaviour of dynamic systems was studied. Following this approach, Lyapunov made groundbreaking contributions to the stability theory in dynamical systems during the late 19th and early twentieth centuries.[7]

Lyapunov's work laid the groundwork for understanding the behaviour of systems over time and whether they converge to a stable state or exhibit instability. Later, his research greatly expanded the control and dynamical systems theory field, which continues to be highly influential in various scientific and engineering disciplines.

Poincaré's ideas provided a powerful framework for understanding and analysing the behaviour of dynamic systems, and control engineers working on aerospace problems found his techniques to be highly valuable. Based on Poincaré's approach, these techniques collectively became known as the *state space approach* in control engineering and rapidly became the standard for control engineers working on aerospace problems.

The concept of state has become central to the entire field of applied dynamical theory. What is fundamental about dynamical systems is that their *past history shapes their current behaviour*. Consequently, it is insufficient to describe dynamical system behaviour solely in terms of *instantaneous* relationships between input and output variables. An additional set of variables, known as *state variables*, is necessary to account for a system's history. The adoption of state-space approaches in the analysis of dynamical and feedback systems led to a more profound exploration of scientific and mathematical challenges in automatic control than ever before. This transition can be considered a crucial step in the evolution of control studies as a mature scientific discipline [21].

3.3.2 *Kalman's Legacy and the Emergence of the Mathematical State-Space Approach to Control Systems Analysis and Design*

Mathematical system theory's emergence as a scientific discipline owes a substantial debt to the visionary insights and research endeavours of Kalman, who played a central role in its formation. The origins of mathematical concepts relevant to control theory can be traced back to the time of James Clerk Maxwell, with roots that extend even further into the past. Also, the pre-World War II era saw the development of graphical methods for analysing feedback systems by many notable figures. These *classical control theories* could be observed as key isolated contributions that solved many practical control problems. However, these approaches did not converge to a cohesive scientific discipline.

During this period, *cybernetics* emerged as a transdisciplinary framework focused on circular causality and feedback mechanisms in adaptive systems. While it successfully unified concepts across biological, mechanical, and social systems through its core principles (feedback loops, homeostasis, and information processing), cybernetics struggled to develop rigorous quantitative foundations. This lack of formal

[7] Refer to Sect. 2.5 for other theoretical background of Lyapunov stability theory.

mathematical structure and its expansive scope that often blurred disciplinary boundaries ultimately limited its evolution into a well-defined scientific discipline despite its profound influence on later fields like control theory, artificial intelligence, and systems biology.

General Systems Theory (GST) was another significant development that will be discussed in Chap. 5. It sought to provide a holistic perspective on complex systems but struggled to gain a firm footing due to its somewhat abstract and loosely defined nature.

Beyond their scientific contributions, classical control theory, cybernetics, and GST collectively underscored the need for a formalised framework to analyse and design engineering systems. These disciplines recognised the modern industrial society's growing dependence on engineered systems and the need for systematic methodologies to address their complexity. While each field made significant conceptual advances, they shared a *common limitation*: none fully developed the mathematical foundations required to transform their insights into a rigorous engineering science.

Cybernetics and GST, despite their profound conceptual frameworks for understanding system organisation and behavior, particularly lacked the mathematical formalism needed to operationalise their principles. Cybernetics' qualitative treatment of feedback mechanisms and GST's abstract system hierarchies required stronger formalisation to enable predictive analysis and synthesis. With its precise notation, deductive logic, and computational tractability, *mathematics* emerged as the essential language for this formalisation. Its capacity to systematically encode relationships between system variables, stability conditions, and hierarchical structures proved indispensable for progressing from descriptive theories to prescriptive engineering methodologies.

As a strong advocate of mathematics rigour, Kalman was very well aware of the shortcomings of the approaches of that period. As an example, later, in [1], he sharply criticises the way the network theory was approached during its development. Kalman suggested that there were difficulties and obstacles in the early stages of tackling the problem. He attributed these difficulties to an "over-reliance on engineering/physical intuition" and a tendency to avoid dealing with mathematical issues. He then argues that the approaches were *naive,* implying that the initial attempts at solving the problem were simplistic or lacked sophistication.

Kalman claims that unresolved mathematical issues were associated with the problem, which were not adequately addressed or ignored. He then proposes that the way forward involves a renewed focus on essential mathematics, much of which has been known since the nineteenth century. Hence, a more rigorous and mathematically grounded approach is necessary for progress. Kalman's paper concludes with the following sentence: "Much remains to be done. But the Promised Land has been sighted from several directions".

Pursuing his prophecy, the word used by one of his harsh critics, Issac Horowitz, Kalman proposed an alternative *state space approach* to control systems analysis and design in the 1960s. The state-space theory was developed in continuation of the publication of work by Wiener and others, including Kalman, on optimal estimation and control. The big success of the state-space approach was to mathematically treat optimality and multivariable control problems in a unified fashion. A task formidable, if not impossible, in the classical control framework. Another key advantage of the state-space approach was its compatibility with the advances in digital computer technology, and the analysis and design methods were easily performed in continuous or discrete time domains.

Dynamical systems were represented in the time domain by *state-space equations*. Performance and stability specifications were readily specified in the time domain, often as a quadratic performance index. Mathematical rigour enabled the designers to derive explicit design equations using analytical formulas, often directly optimising the controllers using a particular performance index.

However, later in the 1980s and 1990s, the frequency and time domain viewpoints were successfully merged to take advantage of both approaches. Specifically, frequency-domain performance and robustness specifications were opted for the uncertainty combat that arose in the control systems community. In the end, the best features of the preceding decades were merged into a powerful, unified framework that will be discussed in Chap. 4 as the postmodern control era.

The state-space methodology rapidly spread throughout the community of control scientists, leading to investigations in various directions within control theory. Novel ideas emerged that were specific to the new theory, and some of these investigations resulted in innovative techniques for addressing long-standing control problems, such as the challenge of *stability*. For a concise overview of the evolution of modern control theory, including the frequency domain approach and various techniques from mathematical system theory such as the structural-digraph approach, polynomial-matrix-domain frequential, and geometric approaches, refer to [22].

The state-space approach gave rise to a series of innovative ideas, starting with exploring relationships between state-space models and transfer function representations. This work pioneered by Gilbert [23] and Kalman [24] is known as the *realisation theory*. Algorithms were developed to derive *minimal-order* state-space dynamical models from given transfer function matrices. This pursuit of *minimality* directly led to the introducing of fundamental structural concepts, namely, *controllability* and *observability* [24, 25]. The validity of these characteristics was tested by derived simple rank conditions, and they became essential notions in Kalman's approach to control systems design in the state-space framework. The majority of control scientists adopted these ideas, though some researchers raised serious questions regarding their implementation. As examined in [26], some reservations were posed about the emphasis placed on controllability and observability within state-variable approaches. It is argued that these concepts had been overstated in their importance for this class of problems. They maintained that transfer function methods could adequately verify these properties when needed, and that nothing essential was lost by omitting them entirely if the synthesis problem was properly framed as one involving parameter uncertainty through transfer function approaches. This highlighted the ongoing debates between state-space and frequency-domain methodologies in control theory.

Willems later reformulated the notions of controllability and observability while studying the axiomatic foundations of system theory. Using a *behavioural-based* approach, he generalised the basic system-theoretic concepts, state controllability, and observability. This reformulation demonstrates that controllability can be characterised without an internal representation and can be defined in terms of system behaviour, providing a broader interpretation of the traditional concept of controllability.

Following the sequence of contributions in state-space theory is the introduction of *module theory* for studying linear systems [27]. This sparked a series of mathematically oriented research activities, exploring the incorporation of polynomials and the associated concepts of polynomial division and greatest common divisors (GCD) of polynomials into control theory. On a rigorous mathematical foundation, the state-space model of a finite-dimensional LTI system and the corresponding map determining its dynamics can be considered a polynomial module. Within this framework, the theory of linear systems becomes equivalent to the theory of polynomial modules. Kalman transformed the study of linear systems into the study of *polynomial maps and matrices*, providing entirely new mathematical tools for controlling and analysing linear systems.

Then, investigations into systems over rings followed, which handled the study of linear dynamic systems where the mathematical structures underlying the system equations are rings rather than the more commonly used fields like real or complex numbers. In a natural extension, analysis and control of *bilinear* and *multilinear systems* followed [28]. Among these contributions, Kalman's work on the positive real lemma [29] provided necessary and sufficient conditions for a given transfer function or state-space model to be *positive real*. This is an essential tool in control theory, particularly when dealing with systems that have real-world physical interpretations, such as electrical circuits, mechanical systems, and many other engineering systems. Also, among Kalman's achievements, we can include the *root location* of polynomials [30] and the study of the connection between *Kronecker invariants* and feedback [31]; the notion of Kronecker invariants was associated with the representation of LTI systems in the context of state-space representations and provided valuable information about the system's properties and behaviour.

Subsequently, a new area of research emerged in system theory, focusing on *families of linear systems* [32], where a group of linear dynamic systems that share certain common characteristics or parameters were studied. These systems may vary within a specific range of parameters, and studying such families allows for the analysis of how parameter changes affect system behaviour.

Families of dynamical systems have influenced the field of systems and control theory. It was recognised that feedback is fundamentally essential in control systems because it arises from the understanding that the plant model is an approximation. In practice, we must engineer solutions accommodating a spectrum of plant variations. The exact description of this *family* depends on the specific problem at hand.

This concept was later extended, particularly in robust control, to encompass families of systems modelled by certain naturally occurring, norm-bounded perturbations of a given nominal plant.

In the early 1970s, Kalman introduced a comprehensive algebraic method to address the challenge of *system parametrisation*. He established a universal parameter space tailored for linear time-invariant systems characterised by fixed state space and input/output dimensions. This pioneering effort laid the groundwork for a powerful algebraic-geometric framework used to examine families of LTI dynamical systems. This approach has had profound consequences within the algebraic systems theory and essentially initiated a distinct field of study in control theory.

Classical dynamical ideas, particularly those related to characteristic *modes of vibration* in linear dynamical systems, were known before the evolution of state-space theory. These ideas were foundational in understanding the behaviour of dynamic systems. Modes of vibration were used to represent the inherent vibrational patterns that systems can exhibit when disturbed from their equilibrium positions. These patterns were shown to be determined by the system's physical properties, such as mass, stiffness, and damping. Modes of vibration were often referred to as *eigenmodes* or *eigenfrequencies*.

Each mode corresponds to a specific eigenfrequency, a characteristic frequency at which the system naturally vibrates when excited. In addition to eigenfrequencies, each mode of vibration was associated with a *mode shape*. The mode shape describes how the different parts of a system move relative to each other during vibration. Modes of vibration are crucial in engineering and modal analysis, as they are used to determine a system's natural frequencies, mode shapes, and damping characteristics. With the advent of state-space theory, these system analysis tools were incorporated into state-space formulation. This transition to state-space representation became a fundamental development in control theory.

The prominent figure who pursued this line of research was H. H. Rosenbrock. He introduced the concept of *modal control*, wherein feedback control involves the manipulation of *characteristic* (*modal*) *frequencies*. This novel idea served as the basis for the development of *pole-shifting* or *pole-placement* (*assignment*) control techniques wherein engineers aim to alter the system's poles (eigenvalues) to achieve desired performance characteristics [33].[8]

Rosenbrock's concept of modal control and the subsequent development of pole-shifting techniques profoundly impacted the field of control theory. These ideas generated a significant body of literature, with references including [34–39].

A key result by M. I. Wonham [35] established that a sufficient condition for controllable systems to allocate their closed-loop characteristic frequencies arbitrarily under feedback (within reasonable constraints dictated by physical considerations) is the accessibility of all system states. This fundamental discovery further emphasised the importance of the state in control theory.

The revival of time-response methods, prompted by these developments, greatly influenced control engineering and played a crucial role in addressing the guidance challenges faced by the space program. State-space theory played a vital role in space studies by providing a mathematical framework for modelling and controlling

[8] This paper was initially released as a private report in 1961 by Constructors John Brown, based in London, England.

dynamic systems, which is crucial for accurate navigation, control, and guidance of spacecraft in complex and ever-changing space environments. It helped ensure the success and safety of space missions. In particular, state-space theory and its related mathematical tools were applied in the Apollo project during the 1960s. This contributed to the broader adoption and development of state-space theory in various engineering fields, solidifying its importance and impact in the following decades. For more details, refer to [2].

3.3.3 The Evolution of Filtering Techniques: From Frequency Response to State-Space Theory

Prior to the advent of Kalman's state-space theory, the filtering problem was handled by the frequency response and statistical approaches. Frequency response analysis was widely used to understand and design control systems and filters. Techniques like Bode plots and Nyquist diagrams were instrumental in characterising the frequency response of systems and were commonly applied to analyse and design filters. Wiener filtering was the dominant statistical approach to filtering. It effectively reduced noise from signals and enhanced their quality, making it a useful tool in various applications, including image and audio processing.

Wiener filtering aimed to minimise the mean square error between the original and filtered signals in the time domain. It did not rely on analysing system behaviour in the frequency domain. Instead, it considered the statistical properties of both the signal and the noise to determine the optimal filter that minimises the error.

Frequency response methods, as traditionally used before Kalman, were primarily focused on the analysis of LTI systems. These methods were not particularly designed for handling nonstationary cases, which involve signals or systems that change over time, and were most effective when applied to stationary signals, where the statistical properties of the signal and the noise remain relatively constant. In cases of nonstationary signals, these approaches did not perform optimally or even appropriately.

The Kalman filter, introduced in the early 1960s, revolutionised signal processing by providing an optimal recursive solution to the linear filtering problem. Its time-domain state-space approach, developed by Kalman and Bucy [40], offered distinct advantages over traditional frequency–response methods, particularly for nonstationary systems where it dynamically recostructed system dynamics with measurement noise.

This framework proved exceptionally versatile, finding immediate applications in aerospace, navigation, and control systems. Beyond its direct practical utility, the Kalman filter fundamentally advanced feedback theory by demonstrating how recursive estimation inherently relies on feedback principles, thereby unifying concepts across filtering and control domains.

The Kalman-Bucy filter emerged as a dynamic model with feedback, illustrating the profound *duality* between feedback control and filtering. Integrating optimal control theory, specifically the Linear-Quadratic-Gaussian (LQG) optimal control approach with the Kalman-Bucy filtering method, revolutionised the state-space treatment of control systems.

The development of the LQG problem and its subsequent incorporation into textbooks became a cornerstone, representing a pivotal moment in the history of control and filtering techniques. A key advantage of the approach was its practical handling and natural extension to multivariable plants, the control of which was a significant issue in the late 1960s and 1970s. Refer to Sect. 10.2 for a confirmation on this point.

The intertwined development of LQG control and filtering techniques in modern control theory stemmed from their shared mathematical foundation in state-space representation. This unified framework enabled the simultaneous treatment of optimal estimation (Kalman filtering) and optimal control (LQG regulation) through analogous formulations of stochastic optimisation in the time domain. The LQG problem aimed to find a control strategy that minimised a quadratic cost function while maintaining system stability. This provided an optimal control action based on the available information. Filtering techniques, such as the Kalman filter, focused on estimating the system's state variables optimally based on noisy measurements. The LQG problem combined these objectives by integrating optimal control and state estimation, ensuring that the most accurate state estimates provide the control strategy. This highlighted the trade-off between control performance and state estimation accuracy [21].

3.3.4 Other Developments

The other major developments in the mathematical system theory are the significant contributions of Pontryagin's maximum principle and Bellman's work on dynamic programming in the context of dynamic optimisation under constraints. Bellman was working in the mathematics department of the RAND Corporation,[9] on a project to determine the allocation of missiles for maximum damage to the targets. This work led to the formulation of the *principle of optimality* and *dynamic programming*.

The introduced Pontryagin's maximum principle was particularly effective in addressing problems with constraints imposed on system inputs. These constraints often reflected real-world limitations, such as resource availability, physical boundaries, or legal regulations. The application of Pontryagin's maximum principle

[9] RAND is an American nonprofit global policy think tank research institute and public sector consulting firm which was formed immediately after World War II to connect military planning with research and development decisions. Its name is adopted from a contraction of the term research and development.

demonstrated the effectiveness and potential inherent in the differential-equation-based mathematical system theory approach. It showed that using differential equations to optimise dynamic systems under constraints effectively found optimal solutions.

On the other hand, Bellman's work on dynamic programming was dedicated to solving dynamic optimisation problems, often in the presence of constraints. Dynamic programming provides a versatile mathematical technique for finding optimal solutions over a sequence of decisions or stages.

Like Pontryagin's principle, Bellman's work recognised the significance of dealing with constraints. It provided a structured framework for addressing dynamic optimisation problems under a variety of constraints, making it an invaluable tool for decision-making in fields ranging from optimal control, efficient motion planning algorithms and trajectory optimisation in robotics, economic and financial decision-making, game theory and strategic decision-making, operations research and management science, and to reinforcement learning, where it serves as a key component in value iteration algorithms, which help agents learn optimal policies through trial-and-error interactions with their environment.

These approaches showcased the remarkable potential of a differential-equation-based approach in finding optimal solutions under real-world limitations, whether in resource management, control systems, or other dynamic processes. These techniques remain vital in various disciplines with constrained optimisation problems.

Bellman emphasised the importance of *state* in formulating and solving problems related to decision-making and control. Bellman's insights revealed the great utility of state variables in modelling and addressing a wide range of complex problems, offering a structured and systematic approach.

Furthermore, Bellman's work paved the way for the in-depth examination of the linear multivariable feedback control problem from a state-based perspective. It was in this context that Kalman made significant contributions by providing a definitive treatment of the linear optimal control problem. Kalman's approach involved using a quadratic form of performance index, which is a critical element in optimising control systems. This marked an essential moment in the development of control theory, leading to more effective and rigorous methods for designing and controlling dynamic systems [41]. Kalman [42] states that "Eventually it should be possible to organise system theory as a rigorous and well-defined discipline. One example of this trend is the author's duality principle relating control and estimation."

This work's key advancement over prior feedback design methods was its direct synthesis procedure, deriving system design algorithmically from problem specifications rather than through conventional trial-and-error approaches. This embodied Kalman's philosophy of *mathematically rigorous abstraction* in control design.

The approach fundamentally transformed design practice by:

- Shifting the focus to performance index selection (the quantitative optimisation metric)
- Replacing heuristic methods with formal mathematical techniques.

Interestingly, Michael Athans placed the origin of what we now call modern control theory in 1956. In September of 1956, an international conference on automatic control was organised by the joint control committee of the VDI and VDE.[10] The conference was held in Heidelberg, Germany, where a group of delegates agreed to form an international organisation to promote progress in automatic control. The International Federation of Automatic Control (IFAC) was officially formed in 1957, and the first conference was held in Moscow in 1960. The presented papers and speakers indicated a change in the direction that had been slowly developing during the 1950s, and it is fitting that Kalman presented his seminal paper *On the General Theory of Control Systems* [25]. This paper clearly showed the duality between the problems of multivariable feedback control and multivariable feedback filtering. Following the Moscow conference, the mathematical state-space approach dominated the control theory for the next two decades.

3.3.5 Achilles Heel of Mathematical System Theory

Modern control theory has had a deep, impressive, and effective effect on the control theory for about two decades. A consequence of this storming start of modern control was a decline in the frequency–response or what we call the classical control methodologies. This is observed in the courses taught, books written, and research literature on automatic control of that era. The apparent successes of modern control, along with the overwhelming interest in time-domain state-space theory on the one hand and the diminishing relevance of classical control on the other, were likely seen by many as the realisation of Kalman's promised land. Nevertheless, the course of events did not entirely align with the envisioned promised land. Some control scientists and engineers began questioning the practical applicability of the mathematically grounded state-space approach to control systems analysis and design. Control engineers gave reports of application failures of modern control in practical systems.

It was soon realised that uncertainty and assumptions were the Achilles' heel of mathematical control theory. Rigorous mathematical proofs were based on exact mathematical models and required theoretical assumptions for the proofs to hold. Mathematical models were shown to be only an approximation of the actual plant and did not necessarily conform to the required assumptions. This can be seen in the views of some prominent control scientists and engineers of that time.

The other issue associated with mathematical control theory designs is the lack of transparency in their design process. Engineering design is inherently a transparent

[10] VDI and VDE are German organisations related to engineering and technology. VDI (Verein Deutscher Ingenieure) is the Association of German Engineers and is an organization that represents engineers in Germany. It is one of the largest technical and scientific associations in Europe and plays a significant role in the field of engineering. Also, VDE (Verband der Elektrotechnik, Elektronik und Informationstechnik) is the Association for Electrical, Electronic, and Information Technologies and is another major German organization. It focuses on electrical engineering, information technology, and related areas. VDE is known for setting standards in the electrical and electronic industries.

process, allowing the design engineer to envision the effects of the decisions made at each stage of the design and how they influence subsequent stages. In most mathematical control design methodologies, the designer provides initial conditions and certain required parameter values. The design process then involves solving several algebraic or differential equations and/or inequalities, with the algorithm ultimately yielding a solution.

The concept of internal stability was introduced to prevent right-half plane pole-zero cancellations during the design process. However, in classical control design methodologies, such as the Bode diagram frequency design approach, the impact of adding a pole or zero can be readily evaluated, enabling the avoidance of cancellations. The issue of *non-transparency* of the state-space approach is addressed in a 1975 paper [26] as *obscurity*. It is asserted that the state-space approach lacks transparency, particularly regarding practical control system implementation. The authors argue that the theoretical framework surrounding eigenvalue assignment via state feedback, whether using observers or not, rests on naive and incomplete assumptions when applied to real-world scenarios. They note that state-variable techniques have failed to achieve their stated objectives in practice.

A key limitation lies in how the state-space formulation obscures critical practical considerations such as sensor noise and loop bandwidth constraints, which are largely neglected in the literature. Additionally, while sensitivity analysis under significant plant parameter uncertainty represents a fundamental challenge, it has received minimal attention. Existing state-variable methods focus narrowly on differential sensitivity functions, but even these results are presented in a manner so convoluted that they often lead to erroneous conclusions. After reviewing topics such as eigenvalue realisation by state feedback, state observer design, decoupling theory, sensitivity theory, and controllability and observability, the authors argue that while certain problems hold substantial mathematical interest, many lack practical relevance for feedback control engineering. In cases where these problems do apply, the state-variable formulation often needlessly complicates the analysis. Transfer function methods, by contrast, consistently yield more evident and more interpretable results for engineering applications.

Zadeh articulates similar concerns through a distinct theoretical perspective, particularly regarding the mathematical formalisation of control systems. He recalls that during the early 1960s, his skepticism toward mathematical formalism represented a minority view. At that time, Lyapunov's stability theory and differential-equation-based nonlinear system analysis dominated control theory research, alongside growing work in optimal control and systems optimisation. Kalman's groundbreaking contributions, particularly his controllability, observability, and filtering frameworks, stood at the forefront of these developments, commanding widespread attention despite Zadeh's reservations about their practical limitations [43]. As will be seen later, Zadeh's doubts were initiated by the practicability of precise mathematical control theory for controlling real-world problems.

Also, it can be argued that with the state-variable formulation of linear feedback control system design, to handle the sensitivity problem, the fundamental design conflicts and tradeoffs, the cost of sensitivity reduction and the relations between the price and the benefits are not transparent, and it can lead to impractical designs from a control engineering design perspective. Indeed, from a control engineering practice point of view, the two key elements impair mathematical control theory are *neglecting uncertainty* and substituting it with *assumptions*, and *neglecting the cost of feedback* in a real trade of cost and benefit in control design.

Optimal control was believed to be one of the milestone achievements of mathematical state-space control theory. However, some control scientists and engineers seriously questioned its effectiveness in solving real control problems. In a 1971 paper [44], Rosenbrock, a prominent control theorist and practitioner, wrote as the abstract of his famous paper, "Optimal control with a quadratic performance index has been suggested as a solution to the problem of regulating an industrial plant in the vicinity of a steady state. It is shown that such control is usually not feasible, and if feasible can have serious defects." Also, in the case of a Kalman-Bucy filter design, it is noted that to make the design attractive for industrial applications, it is necessary to assume known the covariance matrices for an added noise, which usually will be unknown and, in part, will often be fictitious [44].

In 1963, the concept of the *small gain* principle, which later played a vital role in the robust stability criteria, was introduced [45]. Willem's perspective on this methodological approach, as documented in [16], has two key characteristics: its inherent generality and analytical clarity. He observed that the framework's principal advantage lay not merely in its capacity to transcend finite-dimensional state-space constraints but more critically in its ability to articulate system-theoretic principles with uncommon transparency. Rather than abstract mathematical generality alone, this emphasis on intuitive conceptual foundations distinguished the approach from other functional analysis and operator theory techniques.

Also, it was shown that under certain assumptions, optimal single input single output state-feedback control laws had some very strong robustness properties. Extending these results to complicated multivariable systems, it was shown that the results remain valid for each input channel to the plant [46]. However, these nice theoretical properties were shown to vanish in the face of model uncertainties. This was declared in a short but effective paper that studied the robustness of optimal controllers (LQG regulators) under different conditions. The abstract has only three words on the robustness properties, "*There are none*". It is finally remarked, "It may, however, be possible to improve the robustness of a given design by relaxing the optimality" [47]. That is moving from *optimality* to *robustness*.

A harsh critic of the modern control approach in the 1960s and 1970s was the founder of the Quantitative Feedback Theory (QFT), Issac Horowitz. Nearly thirty years later, in [48], the author revisits Kalman's foundational assumption—A fundamental but often implicit assumption in control system design holds that process dynamics exhibit only minor parametric variations during operational deployment, and feedback compensation adequately counteracts these perturbations. However, the control system's performance degrades when confronted with substantial parameter deviations beyond the assumed bounds. He points to Achilles Hill of modern control theory and pictures the general atmosphere in that paper that both classical and modern control theories achieved their design objectives solely for nominal plant parameters, with neither framework developing systematic methods to address plant uncertainty. While sharing identical fundamental goals, modern control theory gained disproportionate academic adoption due to its mathematical formalism, *modernism*, and perceived sophistication, despite this shared theoretical limitation regarding robustness.

In his 1993 book [49], Horowitz points to the impracticability of modern control theory and urges the importance of uncertainty and the key role of feedback to handle it. He writes, "Ph.D. graduates from the most prestigious U.S. universities were convinced that the purpose of feedback was to achieve a desired input–output relation (if they belonged to the classical school), or eigenvalue realisation (if they were 'modernists', which is even more restrictive than the classical objective) for a fixed plant." He acknowledges that even the robustness label for feedback control is unnecessary as uncertainty has always been what is genuinely feedback about. This philosophical approach to control design would automatically reject most of the mathematical control theory.

Horowitz presents what he calls the *fads* in control theory. He believed that "control theory is barren without its true essence of coping with uncertainty." He claims that the host of Academia working in the modern and even postmodern control methodologies, ignoring uncertainty, have leapt from one area to another, and only a few of their results may be useful and relevant, and the rest are mainly misplaced and spurious fads [49]. He asserts that one of these fads is the Lyapunov stability theory in the early 1960s, as it did not consider uncertainty. The other huge fad of the 1960s is the optimal control as it too almost completely neglected uncertainty. He finalises his point by declaring that the greatest fad is that of state space, controllability, observability, LQR, pole-placement with its observers.

Horowitz quotes Kalman in [49], saying that he publicly declared, "The Laplace Transform is dead and buried." He immediately gave the biblical reference from the New Testament that "the stone which the builders have scorned has become the cornerstone," referring to the importance gained by frequency response and Laplace Transform concepts in the later developments of control theory.

This section is concluded by a statement from Zadeh in [43] "Classical systems analysis and control[11] can point with pride to many brilliant successes. But these successes should not obscure the fact that the world is changing, that high machine intelligence is becoming a reality, and that methods that have proved to be so successful in the past may not provide the right tools for addressing the problems of the future."

[11] In this statement classical refers to both the classical and modern control designs discussed in this Chapter.

3.4 Conclusions

This chapter discussed the achievements of control theory from the late 1930s to the early 1960s. This was a time when effective control designs were designed based on the classical control theory, and later, with the advent of modern control theory, rigorous mathematical control theory flourished. The main accomplishments have been frequency design techniques, transfer function-based analysis and design methodologies, and state-space and optimal control methodologies.

It is shown that the World War II push accelerated control systems development theoretically and application-wise in stabilising and guiding aircraft, missiles, and anti-aircraft systems.

It is argued that one of the most fundamental developments was the emergence of the modern control approach to systems and control concepts, where mathematicians and engineers defined and made substantial contributions to the notion of optimality in control theory. Kalman's legacy and impact on mathematical systems theory are highlighted, and then the Achilles Heel of the well-developed modern control techniques are elaborated.

References

1. R. Kalman, Old and new directions of research in system theory, in *Perspectives in Mathematical System Theory, Control, and Signal Processing: A Festschrift in Honor of Yutaka Yamamoto on the Occasion of his 60th Birthday* (Springer, 2010), pp. 3–13.
2. A. MacFarlane, The development of frequency-response methods in automatic control. IEEE Trans. Autom. Control **24**(2), 250–265 (1979)
3. H.M. James, N.B. Nichols, R.S. Phillips, R.S. Phillips, *Theory of Servomechanisms* (McGraw-Hill New York, 1947)
4. H. Nyquist, Regeneration theory. Bell Syst. Tech. J. **11**(1), 126–147 (1932)
5. H.S. Black, Stabilized feedback amplifiers. Bell Syst. Tech. J. **13**(1), 1–18 (1934)
6. H. Harris Jr., The analysis and design of servomechanisms. Trans. Am. Soc. Mech. Eng. **69**(3), 267–279 (1947)
7. C. Bissell, A history of automatic control, in *Springer Handbook of Automation* (2009), pp. 53–69
8. H.W. Bode, *Network Analysis and Feedback Amplifier Design* (D. Van Nostrand Company, Inc., 1945)
9. H.L. Hazen, Theory of servomechanisms. J. Franklin Inst. **218**(3), 279–331 (1934)
10. D.A. Mindell, Engineers, psychologists, and administrators: control systems research in wartime, 1940–45. IEEE Control. Syst. Mag. **15**(4), 91–99 (1995)
11. D.A. Mindell, Automation's finest hour: bell labs and automatic control in world war II. IEEE Control Syst. Mag. **15**(6), 72 (1995)
12. D.A. Mindell, Anti-aircraft fire control and the development of integrated systems at sperry, 1925–40. IEEE Control Syst. Mag. **15**(2), 108–113 (1995)
13. S. Bennett, A brief history of automatic control. IEEE Control. Syst. Mag. **16**(3), 17–25 (1996)
14. S. Bennett, The emergence of a discipline: automatic control 1940–1960. Automatica **12**(2), 113–121 (1976)
15. T. Babcock-Lumish, *Rise of the Machines: A Cybernetic History*, ed. by T. Rid (Oxford University Press, 2016)

16. J.C. Willems, In control, almost from the beginning until the day after tomorrow. Eur. J. Control. **13**(1), 71 (2007)
17. N. Wiener, *Cybernetics or Control and Communication in the Animal and the Machine* (MIT Press, 1948)
18. C.E. Shannon, A mathematical theory of communication. Bell Syst. Tech. J. **27**(3), 379–423 (1948)
19. S. Bennett, *A History of Control Engineering, 1930–1955* (no. 47) (IET, 1993)
20. A.C. Antoulas, *Mathematical System Theory: The Influence of RE Kalman* (Springer Science & Business Media, 2013)
21. A. MacFarlane, The development of frequency-response methods in automatic control [Perspectives]. IEEE Trans. Autom. Control **24**(2), 250–265 (1979)
22. N. Andrei, Modern control theory. Stud. Inf. Control **15**(1), 51 (2006)
23. E.G. Gilbert, Controllability and observability in multivariable control systems. J. Soc. Ind. Appl. Math. Ser. A: Control **1**(2), 128–151 (1963)
24. R.E. Kalman, Mathematical description of linear dynamical systems. J. Soc. Ind. Appl. Math. Ser. A: Control **1**(2), 152–192 (1963)
25. R.E. Kalman, On the general theory of control systems, in *Proceedings First International Conference on Automatic Control, Moscow, USSR* (1960), pp. 481–492
26. I. Horowitz, U. Shaked, Superiority of transfer function over state-variable methods in linear time-invariant feedback system design. IEEE Trans. Autom. Control **20**(1), 84–97 (1975)
27. R. Kalman, Algebraic structure of linear dynamical systems, i. the module of σ. Proc. Natl. Acad. Sci. **54**(6), 1503–1508 (1965)
28. R. Kalman, *Pattern Recognition Properties of Multilinear Response Functions II* (1980), http://pascal-francis.inist.fr/vibad/index.php?action=getRecordDetail&idt=PASCAL8130066203
29. R. Kalman, On a new characterization of linear passive systems (Center for Control Theory, Research Institute for Advanced Studies (RIAS), 1964)
30. R. Kalman, Algebraic characterization of polynomials whose zeros lie in certain algebraic domains. Proc. Natl. Acad. Sci. **64**(3), 818–823 (1969)
31. R. Kaiman, Kronecker invariants and feedback, in *Ordinary Differential Equations* (Elsevier, 1972), pp. 459–471
32. R. Kalman, Algebraic geometric description of the class of linear systems of constant dimension, in *8th Annual Princeton Conference on Information Sciences and Systems*, vol. 3 (Princeton University Princeton, NJ, 1974)
33. H. Rosenbrock, Distinctive problems of process control. Chem. Eng. Prog. **58**(9), 43–50 (1962)
34. M. Gordon-Clark, A novel approach to the control of dynamically unfavorable processes. IEEE Trans. Autom. Control **9**(4), 411–419 (1964)
35. W. Wonham, On pole assignment in multi-input controllable linear systems. IEEE Trans. Autom. Control **12**(6), 660–665 (1967)
36. M. Heymann, W. Wonham, Comments on pole assignment in multi-input controllable linear systems. IEEE Trans. Autom. Control **13**(6), 748–749 (1968)
37. J. Simon, S.K. Mitter, A theory of modal control. Inf. Control. **13**(4), 316–353 (1968)
38. E. Davison, W. Wonham, On pole assignment in multivariable linear systems. IEEE Trans. Autom. Control **13**(6), 747–748 (1968)
39. E. Davison, H. Smith, Pole assignment in linear time-invariant multivariable systems with constant disturbances. Automatica **7**(4), 489–498 (1971)
40. R.E. Kalman, R.S. Bucy, New results in linear filtering and prediction theory. J. Basic Eng. **83**(1), 95–108 (1961)
41. R.E. Kalman, Contributions to the theory of optimal control. Bol. Soc. Mat. Mex. **5**(2), 102–119 (1960)
42. R. Kalman, The theory of optimal control and the calculus of variations. Math. Optim. Tech. **309**, 329 (1963)
43. L.A. Zadeh, The evolution of systems analysis and control: a personal perspective. IEEE Control. Syst. Mag. **16**(3), 95–98 (1996)

44. H. Rosenbrock, P. McMorran, Good, bad, or optimal? IEEE Trans. Autom. Control **16**(6), 552–554 (1971)
45. G. Zames, Functional analysis applied to nonlinear feedback systems. IEEE Trans. Circuit Theory **10**(3), 392–404 (1963)
46. M. Safonov, M. Athans, Gain and phase margin for multiloop LQG regulators. IEEE Trans. Autom. Control **22**(2), 173–179 (1977)
47. J.C. Doyle, Guaranteed margins for LQG regulators. IEEE Trans. Autom. Control **23**(4), 756–757 (1978)
48. I. Horowitz, It was not easy: a personal view. Int. J. Robust Nonlin. Control: IFAC-Affiliated J. **12**(4), 289–293 (2002)
49. I.M. Horowitz, *Quantitative Feedback Design Theory (QFT)*, vol. 1 (QFT Publications, 1993)

Chapter 4
A Concise History of Control Systems: Postmodern Control Era

> *Uncertainty is the only certainty there is, and knowing how to live with insecurity is the only security.*
> *—John Allen Paulos*

4.1 Introduction

As defined and discussed in Chap. 3, modern control theory encompasses control theory techniques founded on solid mathematical grounds. The common feature of these techniques, is the assumption that all inherent plant properties crucial to control theory are precisely defined in a set of mathematical equations with certain assumptions. Consequently, uncertainty is ignored in these methodologies. The state of control theory was summarised by Fuller in 1963 by pointing to three deficiencies [1]. Fuller raises three explicit points. His first point addresses the issue that the conventional linear control theory of the fifties assumed linear time-invariant plant characteristics, whereas the practical plants are mostly time-varying. The second point raised by Fuller is the neglected ubiquitous presence of noise in real-world control applications. His third point refers to a modelling issue. He argues that plants are generally considered as lumped element networks and, therefore, approximately modelled by linear differential equations, but in reality, partial differential equations more accurately describe them. It is then concluded that this can lead to instabilities in feedback systems due to neglected small parasitic delays. Also, it is noticed that in such cases, dynamic programming and the maximum principle, the two admired techniques of that time, can fail in dealing with the important saturation non-linearity when applied to a high-order plant based on a lower-order approximation.

In contrast, postmodern control approaches encompass all control theories that consider uncertainty and aim to address inaccuracies, time-varying, uncertain, and unknown plants. These methodologies can be categorised in the general frameworks of *adaptation*, *robustness*, and *intelligence* approaches. The birth of some of these postmodern control concepts and their roots date back to the flourishing era of mathematical control systems theory.

A. Khaki-Sedigh, *Control Systems: A Historical and Philosophical Perspective*,
https://doi.org/10.1007/978-981-96-7426-8_4

The concept of postmodernism is encapsulated in the quote, "*The idea of the postmodern is that it is a condition of knowledge in which the instability of the object of study becomes a regular principle of work; the point at which our existence as subjects turns into our object of study*," attributed to an anonymous philosopher, brilliantly conveys the essence of postmodernity in the control systems design framework. It highlights the recognition of changing dynamics and the inherent reflexivity within the design process.

Postmodern control systems design involves a reflexive turn where designers acknowledge their influence on the systems they are designing. This recognition prompts a more transparent, self-aware and interactive approach. Postmodern control systems design encourages an openness to multiple perspectives and rejects rigid, universal truths. Different stakeholders may have diverse views on controlling a system, and designers must handle this *plurality* of perspectives. The concept of postmodernism and its implications in control system design are broadly examined in Chap. 9.

This Chapter briefly reviews the historical developments of the three main postmodern control methodologies: robust control, adaptive control, and intelligent control. These control design techniques are labelled postmodern to emphasise their distinction with the mathematical state-space theory of the 1960s and 1970s that inherently assumed the exact representation of real plants by state-space equations or transfer function models.

4.2 Robust Control Historical Developments

This section provides an overview of robust control theory's historical development and evolution, highlighting its origins in the classical sensitivity design period, the emergence of robust control concepts, and the subsequent introduction of various robust control design techniques. The limitations and challenges associated with robust control designs are also presented.

Few authors use *neoclassical control* terminology to describe a set of methods that follow the classical and modern control methodologies and combine these classical and modern control concepts to develop a control methodology with robustness properties and reduced order to handle complex topologies practically. The control design process will be more efficient than the sole classical or modern control approaches. In the neoclassical approach, the problem is formulated in the *spirit* of classical frequency response methods but from the *modern robust control theory perspective.* That is, the classical approach is being used as a starting point or a reference point, but the problem is being re-examined and re-formulated using the tools and concepts of modern robust control theory [2–4].

4.2.1 The Classical Sensitivity Design Period

The first systematic treatment of uncertainties was introduced in 1945 by Bode in his seminal book [5]. Bode considered sufficiently small plant perturbations (i.e., small uncertainties) and provided an analytical framework for studying the behaviour of closed-loop systems in the presence of such perturbations. Horowitz later extended the Bode approach to designing systems with a degree of robustness to account for finite variations in plant parameters [6]. A concise history of robust control is provided in [7].

The period from 1927, marked by the introduction of negative feedback and the concept of large loop gain, through the 1960s, can be referred to as the *classical sensitivity design period*. The primary focus during this era was on shaping the loop for single-input single-output systems to achieve stability, reduce sensitivity, and suppress noise [7].

In mathematical state-space theory, the challenge of dealing with uncertainties in plant systems was largely disregarded. One significant exception was Cruz and Perkins' introduction of the sensitivity comparison matrix in 1964 [8]. This matrix was explicitly designed to analyse Multi Input Multi Output (MIMO) or multivariable systems and served as a valuable analytical tool for comparing the accuracy improvements achieved in closed-loop and open-loop systems. It represented an early effort to extend Single Input Single Output (SISO) sensitivity results to the more complex domain of MIMO systems.

During this era, the problem of achieving precise control in uncertain systems was commonly labelled as the *sensitivity design problem*. A compilation of the key developments in sensitivity theory during the state-variable period can be found in a collection of reprints edited by Cruz [9], which was published in 1973.

4.2.2 The Birth of Robust Control

In many real-world control design problems, the issue of uncertainty extended far beyond addressing small variations in plant parameters, as is typically handled in sensitivity-based design approaches. It necessitated a more intricate mathematical framework, which we now recognise as *robust control theory*. The concept of *robustness* was adopted in the 1970s by theorists like Safonov to describe *uncertainty tolerance* and the term *robust control* was used by Davison in 1975 in a conference paper, and later in a well-known journal paper [10], and also by Pearson and Staats in 1974 [11].

The building blocks of robust control theory are the following theoretical achievements:

- The *small gain* principle introduced by Zames in 1963. This notion played a key role in robust stability criteria [12].

- Kalman's robustness study of optimal state feedback controllers for SISO plants. He proved a closed-loop infinite gain and 60 deg phase margins [13].
- The *Nyquist stability criterion* generalised to multivariable systems by Rosenbrock [14] and MacFarlane and Postlethwaite [15].
- The concept of *coprime matrix fraction description* of multivariable systems introduced by Youla et al. [16] and Desoer et al. [17] in 1976 and 1980, respectively.
- The *parameterisation of all stabilising compensators* introduced by Youla. This was instrumental in achieving robust stabilisation of multivariable systems, a concept now widely referred to as the *Youla parameterisation* [16].

As early as 1963, introducing the small gain principle concept marked a significant step towards robust stability criteria [12]. Studies demonstrated that, under specific assumptions, optimal SISO output state-feedback control laws displayed significant robustness properties [13]. These findings were extended to complex multivariable systems, confirming the applicability of the results to each input channel of the plant [18].

However, as discussed in Sect. 3.3.5, investigations into the robustness properties of optimal controllers (LQG regulators) yielded disappointing results. It was concluded that the robustness of a given design could be enhanced by relaxing the requirement for optimality [19]. This signified a shift from a focus on *optimality* to emphasising *robustness*. This observation distinctly revealed that the *well-embraced*, *well-posed* and *well-solved* optimal control problems in the mathematical system theory context could fail in practical applications.

Many robustness analysis techniques and robust control design strategies have been developed since the mid-1970s. In what follows, some of the key design techniques are briefly represented.

Control theorists in the late 1970s and early 1980s displayed significant interest in optimal LQ (Linear Quadratic) and LQG (Linear Quadratic Gaussian) design methodologies. Later, it became evident that the strong robustness properties of these controllers could be compromised when dealing with state estimation or multiple sources of uncertainty. Nonetheless, a robust multivariable design philosophy emerged through continued effort and ongoing refinements, known as the *LQG/LTR* (Linear-Quadratic-Gaussian/Loop Transfer Recovery) approach [20].

The notion of *Singular Value Decomposition* (SVD) was employed in frequency domain robust control analysis and design techniques in the late 1970s. This attempted to extend the classical Bode design approach to multivariable systems and provide a foundation for a blended classical-modern control approach. However, this approach was overly conservative when evaluating stability and performance in the presence of structured plant variations. To address this issue, Doyle introduced the Structured Singular Value (SSV) concept in 1982 [21]. The SSV of a matrix, denoted as $\mu(M)$, gave rise to what is known as the *μ-synthesis* approach.

Another popular approach to robust control was the $H\infty$ approach, introduced by Zame in 1983 [22]. In addition, Youla and Bongiorno proposed the H2 optimal sensitivity design problem in 1985 [23].

Horowitz and his colleagues pursued another distinct line of research in robust control. They introduced the concept of *templates* to model plant uncertainties and based their design on the Nichols Chart using a *loop-gain shaping* approach. This method, known as *Quantitative Feedback Theory* (QFT), originated in earlier philosophical perspectives on feedback control theory and is presented in [24].

Horowitz's design philosophy emphasises that uncertainty should be inherently embedded into every feedback design, to the extent that even explicitly adding robustness is viewed as *superfluous*. He takes this viewpoint to the extreme, arguing that emphasising robustness in feedback design is akin to stating "live human beings" whenever human beings are discussed [24].

4.2.3 The Limitations of Robust Control

Proponents of robust control often assert that it can effectively address uncertainty. However, it is important to highlight that traditional robust control approaches typically depend on accurate a priori assumptions about the scale and nature of the uncertainty. Unfortunately, this necessary online information is often unavailable or inaccessible to the controller designer. Moreover, attempts to generate this type of online data frequently depend on further a priori assumptions regarding the system.

Assumptions are the consequences of simplifications to derive mathematically feasible analysis and design results. This is one of the limitations of mathematical system theory (MST) that is present in the robust control design approach. Reference [25] refers to this issue by stating that even with the well-developed robust control theories, control designs may not be sufficiently robust, and several failures of robust control in practical implementations are taken as examples of this deficiency. The paper also notes that the main reason for these failures is that at the time of the control system design, the nature and size of model uncertainties were unavailable and, in practical scenarios, could not have been correctly estimated. These robust control design shortcomings led some researchers to seek a revolutionary paradigm shift by abandoning the common belief that 'one must assume something to know something' [26].

Fragility is another issue associated with robust control designs resulting from optimising a cost function. Fragility occurs when small variations in the controller parameters lead to significant changes in the closed-loop system's stability and performance. Fragile controllers often result from optimisation methods that minimise specific performance or robustness metrics, such as H2, H∞, L1, and μ formulations. These controllers are undesirable because they are sensitive to uncertainties and errors in implementation.

The assumption that the designed controller is precisely implemented is implicit in most control system designs, particularly in optimal robust controllers. Although this assumption may hold true and is frequently disregarded, it is important to recognise that controllers operate within feedback systems, and the closed-loop system must maintain robustness against a certain degree of uncertainty in its coefficients. Uncertainties can arise from controller implementation, which may be subject to inherent imprecision in analogue-to-digital and digital-to-analogue conversion, finite word length, finite resolution measuring instruments, and roundoff errors in numerical computations. Additionally, any theoretical or initial control system design, typically conducted on paper or using mathematical models and simulations, may require adjustments during practical implementation. This is because attempting to condense all performance requirements into a single numerical norm value is often inadequate due to the complexity of control systems, necessary trade-offs, and application-specific requirements that can vary widely. This highlights the necessity of ensuring sufficient stability and performance margins around the transfer function coefficients and other defining implementation parameters of the nominal controller.

This issue is addressed in the context of both fragile and optimal controllers, demonstrating that controllers designed on optimisation-based approaches can produce *extremely* fragile controllers. In these cases, even the smallest deviations of the controller's coefficients from the nominal values can *destabilise* the closed-loop control system [27].

Conservatism plays a central role in robustness analysis and robust control design. Designing for worst-case scenarios is a fundamental aspect of robust control design, contributing to its inherent conservatism. This design philosophy can result in high-order controllers even for low-order uncertain plants. High-order systems often exhibit *instability holes* in their stability regions, which are gaps where the system becomes unstable. These holes are a well-known feature of high-order systems. Optimisation algorithms can accidentally place controller parameters close to these holes due to the absence of margin considerations during optimisation. In a sense, this design approach *shifts* sensitivity from the plant to the controller [27].

4.3 Adaptive Control Historical Developments

Adaptive control represents another prominent philosophical school in control theory dedicated to managing uncertainty. One of the earliest papers on adaptive control theory [28] described adaptive control as "an approach to the logical design of a class of servomechanisms which we have here designated as *adaptive*." The concept of adaptation and the term adaptive control originated in biology literature, describing the ability of biological organisms to adjust to their surrounding environment. The authors used it in the same sense in control theory to describe control systems that monitor their performance and adjust some parameters to achieve enhanced closed-loop performance. For a comprehensive exploration of the history of adaptive control, there are several informative references available, including [29–32].

Adaptive control research has a rich and extensive history. There have been debates about the principles and precise definitions of adaptive control. For example, in the discussion following [28], Ian McCausland from the University of Toronto in Canada objects to the term *adaptive* and writes, "The authors define an adaptive control system as one which monitors its own performance and adjusts some of its own parameters in the direction of better performance. However, any feedback control system may be said to monitor its own output or outputs and to adjust something in the system in such a way as to decrease the discrepancy between what the performance is and what it ought to be. To call one system adaptive because it varies parameters and to exclude another system from this category because it only varies signals seems to be an undue restriction of the meaning of the word adaptive." In response, the paper's authors presented their perspective, and it is interesting to note that they concluded their defence statement by identifying some of the terms used to describe their methodology as *input-controlled*, *sympathetic*, *learning*, *self-adjusting*, *self-designing*, and *adaptive*. The authors then concluded that the term adaptive was most suited for their proposed methodology.

Among the many interesting references on adaptive control, a 1961 PhD thesis notes: "In recent years, interest has been developing rapidly in designing control systems capable of adjusting to changing environmental conditions which affect the system's performance. In particular, this interest has centred on designing systems that will automatically sense the need for and carry out the adjustments. Such systems have come to be called *self-adaptive* or more simply *adaptive*" [33].

Commencing in the early 1950s, the quest for autopilot systems for high-performance aircraft ignited a wave of research activity in adaptive control. High-performance aircraft exhibited dramatic shifts in their dynamics when transitioning between different operational states. Autopilots relying on constant-gain, linear feedback mechanisms proved effective within specific operational conditions but were challenged to maintain performance across the entire range of flight conditions. Simultaneously, the process control domain also demanded automated fine-tuning for its controllers.

Whitaker et al. [34] introduced Model Reference Adaptive Control (MRAC) in response to the autopilot control challenge based on what was later called the *MIT rule*. The autopilot design addressed a servo problem, ensuring the system output tracked a desired set-point despite parametric uncertainties. The adaptive mechanism functioned as an outer loop, complementing an inner loop that comprised a conventional feedback control system. Various MRAC schemes were devised with adaptive laws designed using methods such as the sensitivity method and the MIT rule to estimate unknown parameters.

This approach primarily focuses on identifying the uncertain plant and dynamically updating the control design using the latest identified (estimated) model. Although it minimised the conservativeness of robust control design, it faced issues with stability and convergence. The principle that justifies using identified or estimated values as if they were actual values is the *certainty equivalence principle*, detailed in Sect. 5.5.1.

However, employing the certainty equivalence principle requires further assumptions regarding the variables' unconditional expected values or the joint probability distribution. It can be argued that these assumptions are no more restrictive than the linearity assumption with the fixed structure control systems or servomechanisms design, as indicated in [35]. The following timeline can categorise the main historical developments in adaptive control.

4.3.1 The Enthusiasm to Disappointment Era

From the early 1950s to the second half of the 1960s, this period started with much interest in solving the many problems encountered in control engineering through adaptive control strategies. Even a survey paper on adaptive control was published in 1958 [36]. The main adaptive control techniques were the model reference adaptive control (MRAS) approach based on the MIT rule, the self-oscillating adaptive systems (SOAS), and Kalman's self-optimising control systems. The decade ended with the disappointment of adaptive controllers' performance in practice. This is pictured in [37] by characterising the era as 'a lot of enthusiasm, bad hardware and nonexisting theory', which indicates the two main reasons behind the adaptive control failures. The disastrous X-15 adaptive flight control design is studied in [38], and the actual crash of the X-15 aircraft in 1967 is simulated and analysed in [39].

4.3.2 The Era of Development of Theoretical Foundations and Stable Adaptive Control Systems

The mathematical control theory of the early 1950s was insufficient for a deep and effective theoretical understanding of the behaviour of adaptive systems, and general closed-loop stability proofs were impossible to achieve. In the 1960s, some fundamental theoretical concepts were developed. The major influential achievements were the development of state-space theories, the introduction of Lyapunov stability theory in the West, and developments in dynamic programming, dual control, stochastic control and stochastic systems. Another major accomplishment was the development of system identification and parameter estimation techniques that were employed in many adaptive control techniques. These advancements were instrumental in upgrading, reshaping and redefining adaptive control systems.

Perhaps the first application of the Lyapunov theory in MRAC is reported in [40], and later in [41]. This stimulated the research on stable adaptive control and was followed by many papers and dissertations [42]. Based on the earlier works, Parks in 1966 published a paper that showed the instability of MIT rule-based MRAC systems for quite simple systems with certain system inputs [43]. Parks managed to present an alternative Lyapunov-based synthesis method and redesigned the previous adaptive loops with a closed-loop stability guarantee. He also introduced the notion of *positive real systems* and Kalman's lemma for such systems in adaptive control.

The other notable paper on the stable adaptive era is [44], which is a culmination of previous works.

Although their work focused on a specific class of LTI plants, it laid the groundwork for developing rigorous stability proofs in MRAC for broader classes of plant models. Advances in stability theory and progress in control theory during the 1960s deepened the understanding of adaptive control and sparked renewed interest in the field throughout the 1970s.

Meanwhile, simultaneous progress in computers and electronics made it possible to implement complex controllers, such as adaptive systems, sparking increased interest in their practical applications. The 1970s marked a period of significant breakthroughs in adaptive control design. Researchers developed and analysed MRAC schemes using the Lyapunov design approach while employing concepts like positivity and hyperstability to establish a wide range of MRAC schemes with robust stability properties. At the same time, parallel work on discrete-time plants in deterministic and stochastic settings resulted in numerous adaptive control schemes backed by rigorous stability proofs [45].

In this period, "the province of experience and art" in adaptive control was substituted by "sound theories of adaptive control" with the realisation of adaptive systems in the context of self-tuning regulators and the model reference approaches, as stated in [46]. These techniques were used successfully in several applications. In addition, with the advent of cheap realisation possibilities using the advances in microprocessor technology of the 1970s, adaptive control techniques were applied to industrial applications in three areas: aircraft control systems, process control and electrical drives [46]. Towards the end of this era, adaptive control was boosted with well-developed, stable adaptive control schemes with desirable parameter convergence properties and successful applications in several practical examples [46, 47].

4.3.3 *The Robust Adaptive Control Era*

The closed-loop stability proofs and convergence properties of the adaptive control schemes of the 1970s were derived under strict assumptions. However, it was shown in [48] that bounded disturbance and command signals could be found such that the output would be unbounded, and the boundedness assumption of these signals is thus insufficient to guarantee the adaptive system's closed-loop stability. Another fundamental assumption was that the process model had the same structure as the actual process.

The adaptation concept tried to present practical designs in uncertain, time-varying, complicated and complex systems. Unfortunately, most adaptive methodologies, similar to robust control, depend on prior assumptions.

A possible disturbing consequence of the prior assumption, as stated in [26], is that the theoretical system identification or adaptive control results may not conform with a scientific theory by Popper's definition when it is assumed that the 'true' plant is LTI with known order and relative degree and other structural assumptions or the 'true' noise is Gaussian, unknown but bounded with specific structural properties.

Also, some critics argued that adaptive control methodologies faced significant limitations and problems in theory and real-world applications. Two selected examples of such comments are presented as follows. It is stated in [49]: "Much theory was over-sold; experimentation or practical application did not always run according to plan; and practitioners are understandably cautious about the implementation of the technology." It is also mentioned in [50] that "The role of simplified models and the robustness to neglected dynamics were other questions that have also arisen." The assumptions underlying adaptive control models could prove critically flawed in practical applications. This issue was studied by Rohrs in his PhD dissertation, which is known as the Rohrs' counterexample, and is discussed in the following paragraph.

Rohrs showed that systems with guaranteed convergence could be susceptible to *unmodeled dynamics* [51]. Through detailed simulation studies with a basic model reference algorithm, Rohrs showed that systems become unstable if the stringent assumptions in stability proofs are unmet. His work highlighted the impact of unmodeled high-frequency dynamics and high-frequency excitation caused by command signals or disturbances. His studies strongly concluded: "...the adaptive algorithms considered cannot be used for practical adaptive control...". Also, "The practical engineering consequences of the existence of the infinite-gain operator are disastrous", and "none of the adaptive algorithms considered can be used with confidence in a practical control system design because instability will set in with high probability" [51].

Rohrs' counterexample profoundly impacted the adaptive control community and acted as a wake-up call, highlighting the need to address robustness challenges in adaptive control, as noted in [49].

A new set of disputes over the practicality of stable adaptive control schemes now followed the high hopes of the 1970s. The nonrobust behaviour of stable adaptive controllers became a hot topic in the early 1980s adaptive control community. The lack of robustness motivated researchers to explore the causes of instability and devise solutions to address them. By the mid-1980s, a series of redesigns and modifications were introduced and studied, giving rise to the field of *robust adaptive control*. A robust adaptive controller was defined as one that guarantees bounded signals despite reasonable unmodeled dynamics and bounded disturbances, with performance errors remaining within limits proportional to the modeling error.

Averaging theory is a notable example [52]. It builds on the observation that there are two loops in an adaptive system: a *fast* ordinary feedback and a *slow* parameter adjustment loop. This observation was crucial to analysing the behaviour of adaptive control systems, and a large body of theory was generated based on it towards the end of the 1908s. It led to an in-depth study of the rate of change of the adaptation rate, revealing that the previous strategy for calculating the adaptation rate may result in very large values, which are undesirable from a robustness perspective. This consequently led to signal normalising approaches.

Application of *dead-zone* was proposed to ensure bounded solutions in the face of arbitrary bounded disturbances. This modification meant that adaptation would halt when the error fell below a threshold determined by prior knowledge of the disturbance. This assumption was later resolved by further modifications in the adaptive law [32].

Closed-loop instability of stable adaptive techniques in the presence of disturbances and/or unmodeled dynamics was also shown by several examples in [53]. These schemes were subsequently robustified by modifying the adaptive laws. For further details, refer to [54].

Initially, the robustified MRAC schemes ensured the existence of a region of attraction where all signals remained bounded, and the tracking error converged to a small residual set. Later, by incorporating special normalisation signals into the adaptive control laws, the region of attraction expanded to encompass the entire space provided that all modelling error terms, required to be small in the low-frequency range, were bounded from above. Refer to Chap. 54 of [45] for more details.

Unlike the methods described above, which aim to ensure robustness by modifying adaptive laws, alternative efforts focused on leveraging the concept of persistent excitation to guarantee the boundedness of all system signals. It was proven that if the reference input exhibits a sufficiently high degree of persistent excitation relative to the disturbance magnitude, all signals in the adaptive system will remain bounded [32].

Solving the robustness problem in adaptive control also addressed the longstanding challenge of controlling linear plants with unknown and time-varying parameters. By the late 1980s, several groundbreaking results were published in adaptive control, particularly in MRAC for linear plants with time-varying parameters. During the late 1980s and early 1990s, adaptive control research shifted its focus to performance properties and extending the achievements of the 1980s to certain classes of nonlinear plants with unknown parameters. These efforts resulted in new MRAC-type schemes inspired by nonlinear system theory and MRAC schemes with enhanced transient and steady-state performance.

Pursuing practical solutions to implementation challenges gave rise to other adaptive control schemes, such as *multiple models adaptive control*. These approaches introduced new assumptions, relaxing some of the earlier constraints and enhancing applicability to nonlinear systems. This is described in [55].

Robust control and adaptive control methodologies are considered two techniques with different design philosophies that combat plant model uncertainties and variations. However, these can be considered complementary methodologies in a unified design philosophy. This development was provided by employing closed-loop plant model identification techniques. In [55], the philosophical distinction between robust and adaptive control is highlighted. From a robust control perspective, adaptive control is viewed as a method to reduce the uncertainty level of the plant model by using a plant model identifier in closed-loop operation. This reflects the adaptive control philosophy of dynamically adjusting to system changes, emphasising flexibility and real-time learning. In contrast, robust control is philosophically rooted in designing controllers that inherently account for uncertainties, prioritising stability and performance across various conditions. Robust control typically addresses both parameter variations and unstructured model uncertainties, resulting in a more conservative, worst-case design approach.

On the other hand, adaptive control focuses on handling parametric variations through adaptation, aligning with its philosophy of continuous improvement and system-specific optimisation. However, it traditionally leaves unstructured model uncertainties unaddressed, a gap historically filled by robustifying adaptation algorithms. Recently, it has been recognised that robustness to unstructured uncertainties should also be integrated into the underlying linear design, reflecting a philosophical shift toward blending the strengths of both approaches.

The interplay between robust control design, closed-loop identification, and adaptive control is further enriched by the use of supervisory (or model-switching) control, which ensures good adaptation transients in the presence of large parameter variations. This approach utilises a priori information, addressing a long-standing challenge in adaptive control and illustrating a philosophical convergence: integrating prior knowledge with real-time adaptation to achieve robustness and adaptability. Ultimately, these developments underscore the complementary philosophical foundations of robust and adaptive control, robust control's emphasis on stability under uncertainty and adaptive control's focus on dynamic optimisation while highlighting their growing synergy in postmodern control theory.

Therefore, the relationship between robust control, identification, and adaptive control is deeply intertwined. Effective interaction between these components is crucial for achieving robust and adaptive control performance in practical applications.

Despite significant advancements, challenges remain in effectively integrating these concepts, particularly in complex and uncertain environments. Future research directions may focus on developing more sophisticated identification algorithms capable of handling non-linearities, time-varying dynamics, and unstructured uncertainties more effectively.

4.4 Intelligent Control Historical Developments

Intelligent control is a field that combines artificial intelligence and automatic control techniques to design systems that can adapt to changing environments and learn from their own experiences.

The term *intelligent control* was introduced by Fu [56] in 1971. Fu's main contribution was to describe systems that incorporate advanced decision-making and problem-solving capabilities, often inspired by human intelligence. His work was foundational in distinguishing the new intelligent control approach from traditional control systems, which typically rely on mathematical models and deterministic algorithms. Fu clarifies in his paper that the term intelligent control does not imply that other control systems lack intelligence. A more fitting interpretation in Fu's view was that the field of intelligent control systems represented the overlap between automatic control systems and artificial intelligence, highlighting the integration of advanced decision-making and problem-solving capabilities from AI into traditional control frameworks. Fu's paper presented an interesting framework for the concept of intelligent control systems with three examples, i.e., control systems with human operators as controllers, autonomous robot systems and man–machine combined systems.

However, the idea of using intelligent methods for control can be traced back to the early days of cybernetics and artificial neural networks. In [57], heuristics is applied to reinforcement learning control, and [58] applies artificial intelligence techniques to control systems design.

In 1996, *soft computing*[1] was introduced in the control systems terminology [59]. Intelligent control is built on soft computing, a set of methodologies that is the foundation for developing, designing, and deploying intelligent systems. The core elements of soft computing include fuzzy logic, neurocomputing, evolutionary algorithms, and probabilistic reasoning. These approaches are often complementary and are increasingly combined to innovate new techniques for designing intelligent control systems.

An early example is the design of neuro-fuzzy structures. Neuro-fuzzy systems integrated fuzzy logic and neural networks to create a *hybrid* approach that combines fuzzy logic's reasoning capabilities with neural networks' learning capabilities. In [59], Zadeh asserts that a significant shift in systems and control is inevitable, emphasising modern systems' growing complexity and interdependence. He suggested that we are transitioning into an era where intelligent systems are becoming the norm. These systems, characterised by their ability to handle complex, dynamic, and often unpredictable environments, could be utilised in modern systems involving multiple interacting components influenced by a wide range of factors. Zadeh also called for a higher priority to be placed on developing methods that can cope with *imprecision*, *uncertainty*, and *partial truth*.

Some of the main sub-domains of intelligent control are the following fields:

- Neural networks
- Fuzzy logic
- Evolutionary algorithms and random optimisation techniques.

[1] In hard computing, the primary goals of the computations are to achieve high levels of precision and certainty.

4.4.1 Neural Networks

The idea of Neural Networks as a mathematical logic system was developed during the 1940s by neurophysiologist McCulloch and mathematician Pitts to model how neurons in the brain work, and the attempts culminated in their 1943 paper [60].

This paper marked a milestone in the history of cybernetics and was fundamental to developing cognitive science and artificial intelligence Despite dissimilar scientific backgrounds, the scientific collaboration was indeed motivated by simultaneous considerations in *philosophy*, *neurology*, and *mathematics*. For a detailed historical discussion of the seminal McCulloch–Pitts collaboration, refer to [61].

In 1949, Hebb introduced the first learning rule for modification of synaptic (connections between neurons), emphasising that neural pathways strengthen with each use, a concept essential to human learning. He argued that their connection is enhanced when two nerves fire simultaneously [62].

With the advancement of computers and the resulting increase in computational power, numerous papers and books were published in the subsequent decades on topics related to neural networks. However, to highlight some of the most significant achievements, one can refer to the neural network models developed by Bernard Widrow and Marcian Hoff at Stanford University, known as ADALINE and MADALINE. Where ADALINE stands for ADAptive LINear Element, and MADALINE stands for Multiple ADAptive LINear Element. It is believed that MADALINE is the first neural network applied to a real-world problem, using an adaptive filter that eliminates echoes on phone lines [63].

The other major development was the introduction of the Perceptron, one of the earliest forms of neural network models. It was introduced by Frank Rosenblatt in 1957 [64], and he published the Perceptron convergence theorem in 1958 [65]. The Perceptron is a single-layer neural network with a binary output, and Rosenblatt's Perceptron convergence theorem states that the perceptron learning algorithm will converge if the data is linearly separable. The theorem was significant because it demonstrated that the Perceptron could learn to correctly classify linearly separable patterns. This later led to the multi-layer neural networks, giving rise to the development of multi-layer perceptrons (MLPs) and the broader field of *deep learning*. Widrow and Hoff also proposed the trainable Perceptron using the Least Mean Square Algorithm [66].

However, between 1960 and 1980, neural network research experienced a near eclipse due to specific limitations on what a neural network could achieve related to the capabilities of single-layer perceptrons. These limitations were the inability to learn nonlinear functions, the inability to solve the XOR[2] problem (a non-linearly separable problem), and the limited representational power. The XOR problem raised in 1969 is described as the starting point of the *dark age* of neural networks or the *AI winter* [67]. The resurgence of neural network research in the 1980s, driven primarily by the discovery of training methods for multi-layer networks, has played a crucial role in the recent revival of interest in neural networks [68].

[2] XOR is a basic logic function that cannot be represented by a single-layer perceptron.

Backpropagation started the *modern era,* or the *real golden age* of neural networks in the 1980s, and with this mechanism, they could be confidently used in a closed-loop control structure. Backpropagation is introduced in [69] to identify and control dynamical systems using neural networks. This is one of the pioneering papers in adaptive neural controllers.

A breakthrough in neural networks was achieved by introducing deep neural networks in the mid-2000s. The concept of neural networks was transformed into deep learning by including too many hidden layers. This significantly increased the computational power of neural networks. Deep learning plays a significant role in numerous complex technologies, including speech recognition, image recognition, language translation, and intelligent control systems. Specifically, deep learning significantly contributes to control systems by enhancing their modelling precision, adaptability, resilience, and resource management efficiency. These advancements ultimately improve performance, dependability, and adaptability across diverse control system applications.

4.4.2 Fuzzy Logic

In 1962, Zadeh, the father of fuzzy logic, published a paper titled *A Critical View of Our Research in Automatic Control* [70], where he openly criticised the control systems research of that time, dominated by the mathematical systems theory. Zadeh states, "Indeed, there are many observers, including this one, who feel that the status of our research in automatic control leaves much to be desired and that a number of corrective steps must be taken without delay in order to redress serious imbalances and shortcomings in our programs." The main reasons for this claim in Zadeh's words are that "very little basic theoretical research on automatic control is conducted in industry" and he sees this due to the abstract in nature of such research that tends to be regarded as *impractical* or *useless* by the decision making persons in industry. He also adds that the research problems in engineering and automatic control are "artificial theoretical problems which have elegant and publishable solutions rather than with realistic complex problems".

In [68], Zadeh is quoted as stating that the solutions to real-life problems in system analysis and control have often been secondary to the development of mathematical theories. These theories frequently address overly idealised problems that have little relevance or connection to practical, real-world applications. In the 1962 view of Zadeh, the field of mathematical system analysis and control, mainly advocated by Kalman discussed in Chap. 3, has focused on developing mathematical theories for idealised problems rather than finding practical solutions for real-life problems and actual challenges of real-world systems. In [59], Zadeh mentions that his initial scepticism about the effectiveness of highly sophisticated mathematical tools in tackling the complex and poorly defined problems common in systems analysis and control began when he became familiar with and understood Pontryagin's work. He describes the work as *beautiful* but *limited* in its effectiveness in solving realistic problems.

Zadeh's earlier doubts about *classical* mathematical systems theory were reinforced, and he understood that the process of mathematising system theory has its limits. Beyond a certain point, efforts to create precise definitions for inherently imprecise concepts, such as adaptation, robustness, and decentralisation, were bound to fail [59]. Ironically, all this happened while co-authoring a book with Charles Desoer titled *Linear System Theory: The State Space Approach* [71].

Fuzzy sets and later Fuzzy control as one of the key intelligent control techniques grew out of these doubts and incompetencies of what we could now call the *classical mathematical system* theory. It was argued that the problem lay in the *fuzziness* of concepts. Attempts to provide a mathematically sound definition of fuzziness led to the concept of a *fuzzy set* introduced in 1965 [72], which marked the beginning of research on fuzzy sets and, later known as *fuzzy logic*.

The introduction of fuzzy sets was confronted with *scepticism* and *hostility* by many well-known control scientists and engineers [59]. As a highly prominent control scientist, Kalman was one of the severest critics of the newly emerging intelligent control field.

The harsh comments of Kalman in 1972 were quoted in [59] by Zadeh. To provide a structured and logically ordered presentation of Kalman's criticism of Zadeh's work, the following points are classified and organised from his comments:

- *General Critique of Zadeh's Presentation*: Kalman acknowledges that Zadeh's proposals could be *severely, ferociously*, and *even brutally criticised* from a technical standpoint but refrains from doing so in this context. Kalman asks: Is Zadeh presenting important ideas or indulging in wishful thinking? This sets the stage for his critique.
- *Fundamental Objection to Fuzzification*: Kalman argues that the main problem in the systems field is not a lack of analytical methods but the need to develop basic concepts and deep insights into the nature of systems. He suggests that the field requires a foundational framework similar to Newton's laws. Kalman asserts that Zadeh's ideas (fuzzification) have no chance of contributing to solving this fundamental problem.
- *Critique of Fuzzy Concepts*: Kalman categorically denies the existence of fuzzy concepts as scientific constructs. He argues that while people may talk about *fuzzy things*, these are not valid scientific concepts. He points out that scientific progress has been made through precise formulations of discoveries, not through fuzzy or imprecise reasoning.
- *Sociopolitical Context of Fuzzification*: Kalman suggests that the permissive political climate may have influenced Zadeh's enthusiasm for fuzziness in the U.S. at the time, which he describes as *unprecedented permissiveness*. He likens fuzzification to a form of scientific permissiveness, arguing that it leads to socially appealing slogans without the rigour of disciplined scientific work or systematic observation.
- *Rejection of Fuzzification as a Scientific Method*: Kalman states that he cannot conceive of fuzzification as a viable alternative to the scientific method. He adheres

to Hilbert's naive optimism, *Wir wollen wissen: wir werden wissell*,[3] emphasising the importance of precision and rigour in science. Kalman criticises Zadeh for presenting trivial examples where fuzziness might seem tolerable or irrelevant while implying (without formal claims) that his methodology could address deep scientific problems. He argues that there is no evidence that fuzzification can solve nontrivial problems.
- *A challenge to Zadeh*: Kalman challenges Zadeh to demonstrate whether abandoning precise reasoning in favour of fuzzy concepts and algorithms can lead to better solutions. He concludes that there is no evidence to support this claim. Kalman acknowledges that if fuzzification is to solve difficult problems, such success is yet to be seen.

It was, therefore, not an easy way forward for fuzzy to compete with the deep-rooted classical mathematics that dealt with utmost precisions with numbers and equations rather than words and vague concepts. However, the winning card was that precision carried a *cost,* and many real-world problems become *intractable* when precise solutions are sought.

The first implementation of Zadeh's idea was accomplished in 1975 by Mamdani [73], which is also called the *Mamdani fuzzy inference system.* In this pioneering work, Mamdani applied fuzzy logic to control systems, specifically demonstrating its application in controlling a small model steam engine. The fundamental concept behind fuzzy logic control, which allows the representation of linguistic terms and imprecise information systematically, was demonstrated by Mamdani to be particularly suitable for systems where precise mathematical modelling is challenging or impractical.

Following Mamdani's work, the field of fuzzy logic saw rapid development, and its applications expanded into various domains. The subsequent development of fuzzy logic technology has been widely spread across various areas, including consumer products such as washing machines, air conditioners, and other appliances where imprecise or uncertain information needs to be processed for effective control. Additionally, fuzzy logic has found successful applications in high-tech industries, including automotive systems (for engine control and anti-lock braking systems), robotics, and process control.

Finally, an important evolution from the original fuzzy logic proposed by Zadeh, known as Type-1 fuzzy logic, was the development of Type-2 fuzzy logic [74]. However, the idea did not flourish till the late nineties, for a survey and historical developments and applications refer to [75]. In Type-1 Fuzzy Sets, membership functions assign a membership grade between 0 and 1 to each object, indicating its degree of belongingness to the fuzzy set. While Type-1 fuzzy logic effectively handles uncertainties by allowing objects to possess partial membership based on their characteristics, Type-2 fuzzy logic systems extend this concept further. In Type-2 fuzzy logic, the membership grade itself is considered a fuzzy set within the [0, 1] interval, introducing a higher level of uncertainty in membership degrees. Type-2

[3] *We want to know: we will become wise*, is a phrase that conveys a desire for knowledge and wisdom in German.

fuzzy logic systems can capture and model more intricate and layered uncertainties in real-world scenarios by representing the membership grade as a fuzzy set. Type-2 fuzzy logic excels in situations where Type-1 systems struggle due to noise, imprecision, or varying expert opinions. For instance, when dealing with concepts that have multiple interpretations or when experts provide different membership functions, a Type-2 system can accommodate these variations and uncertainties. This enhanced capability to address a broader range of uncertainties makes Type-2 fuzzy logic a powerful tool for tackling complex and uncertain real-world problems.

4.4.3 Evolutionary Algorithms

Evolutionary Algorithms (EAs) are a class of optimisation algorithms inspired by the principles of biological evolution. These algorithms are used to find solutions to optimisation and search problems by mimicking the process of natural selection. The key idea is to iteratively evolve a population of candidate solutions over generations, with the fittest individuals being more likely to pass their genetic information to the next generation. These methods are particularly useful for problems where the objective function is complex, non-convex, or lacks a clear structure.

Evolutionary algorithms have found applications in control systems based on the specific characteristics of the control problem and the optimisation goals. Among the evolutionary algorithms commonly used in control systems are Genetic Algorithms (GAs), Particle Swarm Optimisation (PSO) and Simulated Annealing, which are frequently used for tasks such as tuning controller parameters. Another evolutionary computational approach in control systems design is Genetic Programming (GP), which facilitates symbolic-based nonlinear optimisation. In the GP approach, the Darwinian evolution process simulation involves applying fitness-based selection and genetic operators to a population of parse trees within a designated programming language. What distinguishes GP from the conventional GA lies primarily in its representation scheme. Rather than employing fixed-length strings of genes, GP employs variable-length trees consisting of functions and terminals, offering greater flexibility in creating complex solutions through hierarchical combinations of elements [68].

The historical development of evolutionary algorithms and their application in control spans several decades. GAs are a model or abstraction of biological evolution based on Charles Darwin's theory of natural selection, which John Holland founded in the 1960s. Holland's work, particularly his seminal book *Adaptation in Natural and Artificial Systems*, introduced the concept of GAs as a search and optimisation technique inspired by natural selection [76].

Holland's student, David E. Goldberg, significantly contributed to the development of GAs during the 1970s. Goldberg's research helped establish GAs as a robust optimisation technique. The concept of binary-coded chromosomes, crossover, and mutation operators became central to GAs [77].

In recent years, there has been a focus on hybrid and multi-objective evolutionary algorithms. Hybrid approaches combine evolutionary algorithms with other optimisation techniques, and multi-objective optimisation addresses problems with conflicting objectives.

Evolutionary algorithms have proven to be versatile and effective in solving complex optimisation problems, including those encountered in control applications. The ongoing research and advancements in this field continue to explore new algorithmic variants, hybrid approaches, and applications in emerging areas such as artificial intelligence and robotics.

4.5 Conclusions

Modern control theory was founded on solid mathematical grounds, based on the assumption that all inherent plant properties are precisely known and that a set of mathematical equations with certain assumptions exactly defines the system. The transition from modern to postmodern control approaches is historically reviewed in this Chapter. Postmodern control methodologies are defined as the control analysis and design theories that consider uncertainty, inaccuracies, time-varying and unknown plant challenges. Postmodern control is categorised in the general frameworks of adaptation, robustness, and intelligence approaches. At its core philosophy, postmodernism in control has embraced changing dynamics, and there is an acknowledgement that the systems being controlled are not exactly previously known and modelled. The historical developments of the three main postmodern control methods, robust control, adaptive control, and intelligent control, are reviewed, and their distinction with the mathematical state-space theory of the 1960s and 1970s is highlighted.

References

1. A. Fuller, Directions of research in control. Automatica **1**(4), 289–296 (1963)
2. M.E. Campbell, Neo-classical control of structures, Massachusetts Institute of Technology, 1993
3. F. Assadian, K.R. Mallon, *Robust Control: Youla Parameterization Approach* (John Wiley & Sons, 2022)
4. A. Holohan, M. Safonov, Neoclassical control theory: a functional analysis approach to optimal frequency domain controller synthesis, in *Control and Dynamic Systems*, vol. 50 (Elsevier, 1992), pp. 297–329.
5. H.W. Bode, *Network Analysis and Feedback Amplifier Design* (D. Van Nostrand Company, Inc., 1945)
6. I.M. Horowitz, *Synthesis of Feedback Systems* (Academic Press, 1963)
7. P. Dorato, A historical review of robust control. IEEE Control Syst. Mag. **7**(2), 44–47 (1987)
8. J. Cruz, W. Perkins, A new approach to the sensitivity problem in multivariable feedback system design. IEEE Trans. Autom. Control **9**(3), 216–223 (1964)
9. J. Cruz, *System Sensitivity Analysis* (Dowden, 1973)
10. E. Davison, The robust control of a servomechanism problem for linear time-invariant multivariable systems, IEEE Trans. Autom. Control **21**(1), 25–34 (1976)

11. J. Pearson, P. Staats, Robust controllers for linear regulators. IEEE Trans. Autom. Control **19**(3), 231–234 (1974)
12. G. Zames, Functional analysis applied to nonlinear feedback systems. IEEE Trans. Circuit Theory **10**(3), 392–404 (1963)
13. R.E. Kalman, When is a linear control system optimal? J. Basic Eng. **86**(1), 51–60 (1964)
14. H. Rosenbrock, The stability of multivariable systems. IEEE Trans. Autom. Control **17**(1), 105–107 (1972)
15. A.G. MacFarlane, I. Postlethwaite, The generalized Nyquist stability criterion and multivariable root loci. Int. J. Control **25**(1), 81–127 (1977)
16. D. Youla, H. Jabr, J. Bongiorno, Modern Wiener-Hopf design of optimal controllers–Part II: the multivariable case. IEEE Trans. Autom. Control **21**(3), 319–338 (1976)
17. C. Desoer, R.-W. Liu, J. Murray, R. Saeks, Feedback system design: The fractional representation approach to analysis and synthesis. IEEE Trans. Autom. Control **25**(3), 399–412 (1980)
18. M. Safonov, M. Athans, Gain and phase margin for multiloop LQG regulators. IEEE Trans. Autom. Control **22**(2), 173–179 (1977)
19. J.C. Doyle, Guaranteed margins for LQG regulators. IEEE Tran. Autom. Control **23**(4), 756–757 (1978)
20. J. Doyle, G. Stein, Multivariable feedback design: Concepts for a classical/modern synthesis. IEEE Trans. Autom. **26**(1), 4–16 (1981)
21. J. Doyle, Analysis of feedback systems with structured uncertainties, in *IEE Proceedings D Control Theory and Applications*, vol. 129, no. 6 (Institution of Electrical Engineers, 1982), pp. 242–250
22. G. Zames, B. Francis, Feedback, minimax sensitivity, and optimal robustness. IEEE Trans. Autom. Control **28**(5), 585–601 (1983)
23. D. Youla, J. Bongiorno, A feedback theory of two-degree-of-freedom optimal Wiener-Hopf design. IEEE Trans. Autom. Control **30**(7), 652–665 (1985)
24. I.M. Horowitz, *Quantitative Feedback Design Theory (QFT)*, vol. 1 (QFT Publications, 1993)
25. M.G. Safonov, Origins of robust control: early history and future speculations. Annu. Rev. Control **36**(2), 173–181 (2012)
26. M.G. Safonov, Robust control: fooled by assumptions. Int. J. Robust Nonlinear Control **28**(12), 3667–3677 (2018)
27. L.H. Keel, S.P. Bhattacharyya, Robust, fragile, or optimal? IEEE Trans. Autom. Control **42**(8), 1098–1105 (1997)
28. R. Drenick, R. Shahbender, Adaptive servomechanisms. Trans. Am. Inst. Electr. Eng. Part II: Appl. Ind. **76**(5), 286–292 (1957)
29. A.M. Annaswamy, A.L. Fradkov, A historical perspective of adaptive control and learning. Annu. Rev. Control **52**, 18–41 (2021)
30. W.S. Black, P. Haghi, K.B. Ariyur, Adaptive systems: history, techniques, problems, and perspectives. Systems **2**(4), 606–660 (2014)
31. J. Baillieul, T. Samad, *Encyclopedia of Systems and Control* (Springer, 2021)
32. P.V. Kokotović, *Foundations of Adaptive Control* (Springer, 1991)
33. M.L. Moe, The Use of Time-moments in model-reference adaptive control systems. Ph.D. Thesis, Northwestern University, 1961
34. H.P. Whitaker, J. Yamron, A. Kezer, Design of model-reference adaptive control systems for aircraft, Massachusetts Institute of Technology, Instrumentation Laboratory, 1958
35. H.A. Simon, Dynamic programming under uncertainty with a quadratic criterion function. Econom. J. Econom. Soc. 74–81 (1956)
36. J. Aseltine, A. Mancini, C. Sarture, A survey of adaptive control systems. IRE Trans. Autom. Control **6**(1), 102–108 (1958)
37. K.J. Åström, Theory and applications of adaptive control—a survey. Automatica **19**(5), 471–486 (1983)
38. E. Adkins, L. Taylor Jr, Adaptive flight control systems-pro and con (NASA Flight Research Center, Edwards, California, 1964)

39. Z.T. Dydek, A.M. Annaswamy, E. Lavretsky, Adaptive control and the NASA X-15 program: a concise history, lessons learned, and a provably correct design, in *2008 American Control Conference* (IEEE, 2008), pp. 2957–2962
40. L.P. Grayson, The design of nonlinear and adaptive systems via lyapunov's second method, in *Presented at the Joint Automatic Control Conference* (Polytechnic Institute of Brooklyn, 1963)
41. L.P. Grayson, Design via Lyapunov's second method, in *Joint Automatic Control Conference*, no. 1 (1963), pp. 589–595
42. J.W. Graham, A study of a model-referenced adaptive control system, 1968
43. P. Parks, Liapunov redesign of model reference adaptive control systems. IEEE Trans. Autom. Control **11**(3), 362–367 (1966)
44. B. Shackcloth, Design of model reference control systems using a Lyapunov synthesis technique, in *Proceedings of the Institution of Electrical Engineers*, vol. 114, no. 2 (IET, 1967), pp. 299–302
45. W.S. Levine, *The Control Handbook (three volume set)* (CRC press, 2018)
46. T. Soderstrom, *Methods and Applications in Adaptive Control* (Springer, 1980)
47. C. Harris, S. Billings, Self-tuning and adaptive control: theory and applications, in *IEE Control Engineering Series,* vol. 15 (Peter Peregrinus, Ltd., Stevenage, Herts., England, 1981)
48. B. Egardt, *Stability of Adaptive Controllers* (Springer, 1979)
49. B.D. Anderson, A. Dehghani, Challenges of adaptive control–past, permanent and future. Annu. Rev. Control **32**(2), 123–135 (2008)
50. K.J. Astrom, Adaptive control around 1960. IEEE Control Syst. Mag. **16**(3), 44–49 (1996)
51. C.E. Rohrs, Adaptive control in the presence of unmodeled dynamics, NASA Center for Aerospace Information, 1982
52. K.J. Åström, B. Wittenmark, *Adaptive Control* (Courier Corporation, 2013)
53. P.A. Ioannou, P.V. Kokotovic, *Adaptive Systems with Reduced Models* (Springer, 1983)
54. P.A. Ioannou, J. Sun, *Robust Adaptive Control* (PTR Prentice-Hall Upper Saddle River, NJ, 1996)
55. I. Landau, From robust control to adaptive control. Control Eng. Pract. **7**(9), 1113–1124 (1999)
56. K. Fu, Learning control systems and intelligent control systems: an intersection of artifical intelligence and automatic control. IEEE Trans. Autom. Control **16**(1), 70–72 (1971)
57. M. Waltz, K. Fu, A heuristic approach to reinforcement learning control systems. IEEE Trans. Autom. Control **10**(4), 390–398 (1965)
58. J.M. Mendel, J.J. Zapalac, The application of techniques of artificial intelligence to control system design, in *Advances in Control Systems*, vol. 6 (Elsevier, 1968), pp. 1–94
59. L.A. Zadeh, The evolution of systems analysis and control: a personal perspective. IEEE Control Syst. Mag. **16**(3), 95–98 (1996)
60. W.S. McCulloch, W. Pitts, A logical calculus of the ideas immanent in nervous activity. Bull. Mathem. Biophys. **5**, 115–133 (1943)
61. T.H. Abraham, (Physio) logical circuits: the intellectual origins of the McCulloch–pitts neural networks. J. Hist. Behav. Sci. **38**(1), 3–25 (2002)
62. D.O. Hebb, *The Organization of Behavior: A Neuropsychological Theory* (John Wiley and Sons, 1949)
63. https://cs.stanford.edu/people/eroberts/courses/soco/projects/neural-networks/History/history1.html. Accessed 24 Dec 2024.
64. F. Rosenblatt, *The Perceptron, a Perceiving and Recognizing Automaton Project Para* (Cornell Aeronautical Laboratory, 1957)
65. F. Rosenblatt, The perceptron: a probabilistic model for information storage and organization in the brain. Psychol. Rev. **65**(6), 386 (1958)
66. B. Widrow, M.A. Lehr, 30 years of adaptive neural networks: perceptron, madaline, and backpropagation. Proceed. IEEE **78**(9), 1415–1442 (1990)
67. https://sefiks.com/2017/10/14/evolution-of-neural-networks/. Accessed 23 Ap 2024
68. A. Zilouchian, M. Jamshidi, *Intelligent Control Systems Using Soft Computing Methodologies* (CRC press, 2001)

69. S.N. Kumpati, P. Kannan, Identification and control of dynamical systems using neural networks. IEEE Trans. Neural Netw. **1**(1), 4–27 (1990)
70. L. Zadeh, A critical view of our research in automatic control. IRE Trans. Autom. Control **7**(3), 74–75 (1962)
71. L. Zadeh, C. Desoer, *Linear System Theory: The State Space Approach* (1963)
72. L.A. Zadeh, Fuzzy sets. Inf. Control **8**(3), 338–353 (1965)
73. E.H. Mamdani, S. Assilian, An experiment in linguistic synthesis with a fuzzy logic controller. Int. J. Man-Mach. Stud. **7**(1), 1–13 (1975)
74. L.A. Zadeh, The concept of a linguistic variable and its application to approximate reasoning—I. Inf. Sci. **8**(3), 199–249 (1975)
75. K. Mittal, A. Jain, K.S. Vaisla, O. Castillo, J. Kacprzyk, A comprehensive review on type 2 fuzzy logic applications: past, present and future. Eng. Appl. Artif. Intell. **95**, 103916 (2020)
76. J.H. Holland, *Adaptation in Natural and Artificial Systems: An Introductory Analysis with Applications to Biology, Control, and Artificial Intelligence* (Ann Arbor: University of Michigan Press, 1975)
77. D.E. Goldberg, *Genetic Algorithms in Search, Optimization and Machine Learning* (Addison-Wesley Longman, Boston, Mass, USA, 1989)

Chapter 5
Harnessing Models for Control Systems Analysis and Design

A model is a lie that helps you see the truth.
—Howard Skipper

5.1 Introduction

In natural and man-made systems, the ubiquitous application of *feedback mechanisms* vividly stands out. Whether occurring in the natural world or crafted by engineers, these systems share a fundamental reliance on feedback to enhance comprehension and functionality. Natural sciences are dedicated to studying laws governing the natural world. Similarly, engineering sciences utilise the principles exposed by natural sciences to innovate, design, and maintain artificial systems. The feedback mechanism emerges as a unifying thread in both contexts, pivotal in understanding complex physical systems.

This shared reliance on feedback provides invaluable insights and practical tools for managing the complexities of both natural phenomena and complex man-made systems. Thus, the common application of feedback serves as the connection, bridging the gap between natural and engineering sciences and offering a holistic approach to comprehending and manipulating the world around us.

Modern feedback control[1] has three main elements:

- System modelling
- System analysis
- Control system design

The primitive feedback control in the Industrial Revolution and before that was concerned with *heuristic-artistic-innovation case-based* designs with no *modelling-based* analysis. The applications were simple plant regulations, and the lack of modelling and analysis did not affect the feedback design process.

[1] This refers to the feedback control theory after the introduction of modelling and system theory concepts.

A. Khaki-Sedigh, *Control Systems: A Historical and Philosophical Perspective*,
https://doi.org/10.1007/978-981-96-7426-8_5

This feedback design ecosystem persisted until the late nineteenth and early twentieth centuries, with few exceptions. Its main prevailing philosophical and scientific ideas can be summarised as follows:

- Feedback designs were often practised as a craft passed down through generations of artisans and skilled engineers of the particular field. In such cases, *practical knowledge* and *hands-on* experience were highly valued in feedback designs.
- Control feedback designs were approached with *empirical* problem-solving techniques. This relied on trial and error, observation, and practical experience to devise feedback loops rather than employing systematic scientific methods.
- Feedback designs were often intertwined with *natural philosophy*, and designers relied on the principles and ideas of natural philosophers to design and construct artefacts with inherent feedback loops.
- In the very early stages and before the Industrial Revolution, feedback design achievements were mainly recognised for their *aesthetic* value and the *magical* scenes they provided, blending functionality and beauty.
- A common characteristic of feedback designs was their trial-and-error nature. Only *incremental improvements* were made based on the performance and functionality of their artefacts.
- The achievements in feedback design were often *localised*, and designers of feedback loops in various disciplines frequently had to begin anew.

However, as industrial feedback control systems became more widespread in the late nineteenth and early twentieth centuries, designers faced pressing theoretical and practical challenges that demanded immediate attention. Two key concepts emerged and were developed to help solve the anomalies at that time. These were the introduction of *mathematical modelling* to describe the behaviour of physical systems and the noble concept of *system theory*.

The introduction of mathematical modelling enabled advanced mathematical tools for analysing and designing control systems. Models have been developed to offer a parametric system representation designed to encapsulate real-world systems' essential characteristics by incorporating adjustable parameters. These parameters serve as variables or factors that can be tuned to reflect the complex aspects of the underlying systems. Additionally, using models within a system theory framework enabled control system designers from various engineering fields to adopt a *unified symbolic representation* easily understood by all. This facilitated a mutual theoretical understanding and provided a common platform for discussion. Furthermore, the analysis and design methodologies became readily available and easily transferable to all engineering disciplines.

This chapter presents the modelling aspects of control systems design, while the next chapter introduces the concept of systems theory. There are several main approaches to modelling for control systems analysis and design, and they can be categorised in different ways. These approaches vary in their level of structural and parametric complexities, mathematical formulation, application areas, and the assumptions they make. There are different classifications for modelling approaches. In this chapter, the following general approaches are presented:

- Analytical mathematical modelling
- Modelling from data

5.2 Analytical Mathematical Modelling Approach

Analytical models, rooted in physical principles or mathematical equations, can offer *precise* or *approximate* solutions to the governing equations of a system. This capability allows analytical models to be effective tools for understanding, simulating, predicting, and optimising the behaviour of systems. The widely used analytical models are state-space equations, transfer function, and input–output models, as well as their extensions and modifications. After introducing analytical mathematical models, many *model characteristics* were established. They exhibited key features that define a dynamical system, including pole-zero configurations, order, and structural properties. Additionally, these models are employed to *estimate* and *predict* system performance and are utilised for model-based closed-loop *control system design* and *optimisation.*

While mathematical models are invaluable in control system analysis and design, compelling arguments exist against their application in certain scenarios. Take, for instance, systems with intricate physics, as evident in biomechanics or soft robotics. In biomechanics research, creating accurate models to represent the complexity of human movement is a demanding task. The intricate physics involved in the interactions between various anatomical components makes it challenging to devise a comprehensive mathematical model. While simplified models may provide some insights, they often fall short of capturing the full intricacies of the human body's biomechanical complexities. Also, soft robotics, often characterised by deformable structures and complicated interactions, pose a formidable challenge for conventional modelling approaches. Soft robotic components' complex and dynamic nature may defy straightforward encapsulation within traditional mathematical frameworks.

Similarly, consider systems exposed to complex disturbances, as exemplified by the variable and often turbulent nature of ocean currents that affect the autonomous underwater vehicles' trajectory and performance or in power grids that incorporate variable renewable energy sources, such as wind and solar, encountering complicated disturbances due to the variability and intermittency of these sources. Such unpredictable and complicated disturbances can lead to highly complex models, making capturing and accurately predicting the system's behaviour challenging. Despite the availability of first-principle models, their complexity can make them impractical for control design.

Moreover, systems rely on sophisticated ways or modes through which they perceive and gather information from their surroundings. In control systems, robotics, and automation, sensing modalities encompass various sensing and data acquisition methods. These modalities provide distinct types of information about the environment, allowing systems to make informed decisions and respond appropriately to changing conditions.

In complicated scenarios, conventional modelling approaches may face limitations. Therefore, there is a pragmatic argument for exploring alternative methods. Rather than relying solely on elaborate mathematical models, a more direct approach, such as empirical or data-driven methods, might prove beneficial. By directly observing and learning from the system's behaviour through data, one can potentially overcome the challenges posed by the complexity and dynamics inherent in complicated systems. In some cases, even if comprehensive first-principle models are available, their complexity may render them impractical for control design [1].

5.3 Modelling from Data

In this section, the concept of modelling from data is categorised as the *classical systems identification* approaches and the more recent *data-driven models*.

5.3.1 System Identification Principles from a General Perspective

Zadeh introduced the term identification in [2] where he states that "The problem of identifying a black box-that is, determining its input–output relationships by experimental means-occurs under different guises in various branches of science. Some writers refer to it as the *characterisation* problem. Others term it the *measurement* or the evaluation problem. Some-borrowing a term from physics-call it *gedanken experiments*.[2] In this paper, we use the term *identification* because it seems to state the crux of the problem with greater clarity than the more standard terms cited above."

A broadly based system-theoretic foundation for the problem of system identification and relevant philosophical discussions are given in [3]. System identification entails several key features from a general system theory viewpoint. These features are summarised as follows [4].

Feature 1. The pragmatic engineering approach to identification involves using a predefined class of potential models established before the identification process begins. Philosophically, this approach prioritises ontology, the study of being or existence, over epistemology, the study of knowledge, which may lead practitioners to overlook certain philosophical concerns. The core philosophical conflict lies between pure epistemology, which views knowledge as the fundamental basis of experience and dismisses speculative metaphysical assumptions about being, and ontology, which asserts that acquiring knowledge requires presupposing an inherent nature of

[2] *Gedanken experiment* is a German term that translates to *thought experiment* in English. It refers to a theoretical experiment that is conducted in the imagination rather than in a physical setting. These experiments are often employed in philosophy, physics, and other scientific disciplines to explore and understand complex concepts or theoretical scenarios.

reality. Pure epistemology treats knowledge as the foundational material shaping our understanding, while ontology suggests that reality possesses independent characteristics that influence what we can know. This creates a fundamental conflict: whether knowledge alone constructs our understanding or whether pre-existing aspects of reality shape that knowledge. Additionally, concerns have been raised about how the generality of models is evaluated, especially in automated systems, where flawed assessments may lead to misinterpretations or misapplications. Properly assessing *generality* is essential in system identification, as it helps determine the suitability and limitations of model classes for specific applications, highlighting broader questions about the interplay between knowledge and objective reality.

Feature 2. The identification techniques typically concentrate on gathering data and selecting the best-fitting model, often overlooking the importance of interacting with the system during the process. In contrast, adaptive system identification methods actively test hypotheses by engaging with the system. These hypotheses, which may concern the system's structure, dynamics, or interactions, act as initial assumptions guiding model selection. Through iterative interaction, researchers collect additional data that either validates or challenges these hypotheses, refining their understanding of the system and improving model accuracy. This approach is especially valuable in cases where results are ambiguous, as it helps narrow down potential models and highlights key behavioral patterns that distinguish between them. By incorporating interaction, adaptive techniques enable a more dynamic and precise system identification process.

Feature 3. The identification process is not conducted in isolation but is typically undertaken for a specific purpose. The purpose itself becomes an integral part of the identification problem. For instance, when evaluating identification, a common criterion is prediction. The primary goal of identification in complex systems often goes beyond pure prediction. Instead, the emphasis is placed on identification for control. In other words, understanding and characterising the system through identification are driven by the need to influence or control the system's behaviour. This shift in focus acknowledges that practical limitations might constrain the capacity for precise prediction of all system aspects in many real-world scenarios.

Furthermore, the significant aspect of human control in complex systems lies in achieving effective control based on relatively basic or incomplete knowledge.

Feature 4. The identification process is challenging as it may sometimes clash with other control objectives or goals. One such conflict may arise from varying requirements for prediction. An identification scheme that works optimally for one type of prediction may perform poorly for another and directly contradict it. A deeper challenge arises when identification is primarily aimed at control. A classic example is the two-armed bandit problem, where the benefits of gaining knowledge through identification must be carefully balanced against the costs of sub-optimal control. In situations where the goal is to control, the trade-off between gaining knowledge and achieving optimal control becomes a complex and delicate balance.

Feature 5. The identification problem often struggles for precise solutions as it demands a model's response perfectly match real measured data, a requirement that loses meaning for non-deterministic systems. Real-world systems are also influenced by noise and uncertainty, making strict determinism impractical. Instead, a more viable approach evaluates how well the model approximates observed behaviour. This method introduces an order relation, prioritising the model's closeness to actual system performance over exact replication. Moving away from strict identity, this approach acknowledges the complexities of non-determinism and the need for practical measures of model accuracy.

Feature 6. The identification problem often presents multiple valid solutions, even in deterministic systems where models can perfectly fit the observed data. However, these models are not equally preferable in practice. When several models explain the data equally well, the simplest, least complex option is the most desirable solution. The preference for simplicity creates an implicit grading among otherwise equally accurate models and aligns with the principle of *Occam's Razor*, which suggests that all other things being equal, the simplest explanation is usually the best. This principle is also known as the *principle of parsimony*.

This principle is commonly used in science, logic, and critical thinking to avoid unnecessary complexity in explanations or theories. By emphasising simplicity, it is acknowledged that more complex models can be overly detailed and may not necessarily provide a better fit to the data. In contrast, simpler models can be more robust and easier to understand, making them more practical and useful for practical applications.

Further insights into the philosophy of science and system identification. There are other obvious and important links between system identification and problems raised in the philosophy of science. *Carnap's Confirmation Theory*, also known as *Inductive Logic*, is a philosophical framework that provides a systematic and quantitative approach to evaluating the degree of confirmation or support that empirical evidence offers for a given hypothesis or theory. Rudolf Carnap developed this framework to clarify the logical relationship between statements, specifically between hypotheses and evidence.

The theory focuses on inductive logic, which involves reasoning from specific observations to more general conclusions. This approach differs from deductive logic, which deals with logical deductions from general premises to specific conclusions. Using inductive logic, Carnap's Confirmation Theory enables researchers to evaluate the degree of confirmation or support for a given hypothesis or theory based on empirical evidence.

In system identification, Carnap's Confirmation Theory provides a useful framework for evaluating the degree of confirmation or support for a given model or hypothesis. By applying the theory to available data, researchers can determine how well a given model or hypothesis is supported by empirical evidence. This approach can help identify the most plausible model or hypothesis, which is essential in system identification.

Also, two other notable concepts from the philosophy of science that provide a rich framework for understanding the principles and mechanisms underlying system identification are Popper's falsification and the Kuhanian scientific revolution. Popper's falsification theory can be seen as a direct application of updating the model set based on new data. When new data arrives, the model that is no longer the best fit (i.e., the one with the highest degree of approximation) is removed from the set of admissible models, thereby falsifying the theory. This corresponds to Popper's idea that a scientific theory can be disproven by empirical evidence.

Similarly, Kuhnian scientific revolutions can be viewed as changes in the complexity ordering of models. This could involve introducing new models or updating existing ones to capture the underlying behaviour of the system better. Kuhnian scientific revolutions are characterised by a shift in paradigm, where a new set of assumptions and concepts is introduced to replace an old one. This could be achieved in system identification by changing the complexity ordering on models, capturing more complex behaviour while maintaining simplicity for previously understood behaviour.

By drawing parallels between these philosophical concepts and system identification, we can better understand the underlying principles and mechanisms that govern both fields. This can ultimately lead to more effective and efficient approaches to modelling complex systems [4]. For Popper's Falsification theory, Chap. 7 and the Kuhnian scientific revolutions concepts are discussed in Chap. 8.

5.3.2 Data-Driven Modelling

Data-driven modelling approaches can be broadly categorised into two main types: *Soft computing models* and *pseudo-models*.

Soft computing models refer to a wide class of techniques utilised to capture and approximate the input–output relationships of complex systems based on measured data or experiences. The main soft computing models include neural networks, fuzzy logic, evolutionary and random optimisation techniques, rough sets, Bayesian networks, and support vector machines. Soft computing models are primarily used for simulating and predicting the behaviour of physical systems. They are widely employed in control systems design and are often integrated into soft computing-based controller design approaches, although they can also be applied in some conventional model-based techniques.

On the other hand, pseudo-models are utilised in data-driven control techniques, such as model-free adaptive control approaches. These models rely on *dynamic linearisation* (DL) data model techniques, which construct an equivalent linearisation data model of the original non-linear system at each operation point of the closed-loop system. A key concept in this approach is the *pseudo-partial derivative* (PPD), a virtual equivalent linearisation data model that estimates the time-varying PPD at each sampling time using measured input–output plant data. The system's dynamic behaviour is captured by observing how changes in the current input signal lead to variations in the output signal.

Additionally, *pseudo order* is defined to enhance the data models for handling complex non-linear unknown systems. The philosophy underlying the concept of dynamic linearisation data models is to provide a data model *specifically* for controller design.

These models are not typically used for system analysis or prediction purposes. Instead, they provide a virtual description of the closed-loop input–output data to facilitate data-driven control techniques. Unlike traditional models (e.g., mathematical or soft computing models), which serve as a foundation for multiple tasks (e.g., system analysis, prediction, and control design), pseudo-models are more specialised.

In [3], it is noted that if a model structure is designed solely to account for the observed behaviour on which it is built, it may fall into the common pitfall of deductive inference. Specifically, such a model risks being inferentially empty, essentially providing nothing more than a rephrased description of the original observations without offering deeper insights or predictive power. Hence, it would suffer from the lack of deductive reasoning power or meaningful generality. It makes the model devoid of significant inference and does not provide any new insights beyond what is already observed. However, as previously mentioned, the recently derived data models encompass a variety that intentionally refrains from asserting deductive reasoning power. A notable example is the *compact form dynamic realisation models* that purposefully serve as pseudo-models, contributing to effective control designs without relying on deductive power [5].

It is worth noting that sometimes, a simplified or approximate representation of a system or phenomenon is also called a pseudo model in the literature. First Order Plus Dead Time (FOPDT) models used in many practical applications are an example. Such models aim to capture the essential behaviour or features while disregarding certain complexities or details. They can provide insights or approximate results without explicitly considering the real system's intricacies. This concept is different from the data-driven type pseudo-models.

In a different framework, when dealing with increasingly complex physical processes we face in the current and future societal challenges for control systems, as discussed in Sect. 10.9, it becomes crucial to understand these systems to improve prediction and control deeply. Handling high-dimensional data, common in these complex processes poses new challenges for researchers and practitioners. Also, the growing interaction between various disciplines, such as system identification, machine learning, and Bayesian statistics, has led to advancements rooted in the *theory of regularisation.* One significant development in this context is using deep neural networks, a hierarchical structure, for learning dynamic models from data. Traditionally, the model's ability to fit the data is balanced without being too complex to be influenced by noise. However, in complex architectures, like deep neural networks, data is fitted in a complex structure and still performs well on validation data, challenging our understanding of this trade-off in the context of modern, sophisticated models. For a survey of deep learning from a system identification perspective, refer to [6].

5.4 Philosophical Considerations in Modelling for Feedback Control

The challenge of deducing the configuration or structure of a system based on observations of its behaviour is an age-old issue with extensive implications. This encounter with data/observation for feature or model extraction is very appealing in engineering and has raised serious philosophical questions concerning the essence of reality and our capacity to infer it from sensory experiences.

Plato's well-known analogy from around 380 BC is a notable reference in this context. In this metaphor, prisoners confined within a cave, like ourselves, can only perceive shadows, either their own or those cast by one another, projected by a fire onto the cave's opposite wall. This analogy is a powerful representation of the inherent uncertainty and fragile nature of the processes of inference that we routinely undertake and often take for granted. It underlines the challenges and limitations inherent in our attempts to understand and interpret the world based on available information. In the context of mathematical modelling, the challenge often involves constructing a representation of a system based on observed behaviours or data. Just as the prisoners in Plato's cave can only see shadows and must infer the reality behind those shadows, modellers work with observed behaviours and data to infer a system's underlying mechanisms. Mathematical models, like the shadows on the cave wall, are abstractions that represent our understanding of complex systems. Constructing these models involves making assumptions, simplifications, and interpretations based on available observations. However, similar to the inherent uncertainty and fragile nature of the prisoners' inferences, mathematical models are approximations of reality and may not capture all the complexities of a system.

In adopting a more practical stance, the validity of models deduced from observing a system's behaviour brings forth profound philosophical issues. This practical perspective, concentrated on evaluating the actual effectiveness and utility of these models, does not alleviate the fundamental philosophical difficulties.

Real philosophical concerns arise when the derived models are utilised for prediction, with the flexibility to incorporate new data as it becomes available. The dynamic models can be updated or refined based on additional observations, which is important for improving their predictive accuracy and reliability. Some significant challenges or obstacles question the ability to validate inductive inference, which is a fundamental aspect of developing predictive models. The reliability of using inductive reasoning to generalise from observed instances to make predictions is questionable. This scepticism about the validity of inductive inference is emphasised as crucial because it is a key component in creating predictive models. Hence, generalising from observed instances to make predictions cannot be conclusively justified, either deductively or inductively [3]. This issue is further elaborated in the following discussions.

5.4.1 Philosophical Foundations of Modelling

Although the modelling approaches may radically vary in concept and even the philosophy of their function, a unified philosophical foundation can be observed in substituting models for real systems. Philosophically, *induction*, *reductionism*, and *mechanism* form the primary foundations and justifications for replacing real physical phenomena with various model forms. Induction allows for the derivation of general models from specific observations, reductionism supports the simplification of complex systems into manageable components, and mechanism provides the basis for explaining phenomena through underlying processes or laws [7–9].

Following Bacon's description of scientific methodology, model-based control system design techniques involve deriving physical facts and laws governing dynamical systems, along with input–output observations and other available measurements, to formulate a model of the system. This model is then validated through simulations and experiments to ensure its accuracy. The process of deriving the model from specific observations and physical principles can be considered induction-based, as it generalises from empirical data to create a broader system representation [9].

Also, *reductionism*, instead of holism, is a philosophical approach that attempts to explain systems and complex phenomena by reducing them to simpler or more fundamental components, such as basic physical laws. It is a theoretical perspective proposing that complex systems, theories, or entities can be comprehended, explained, or analysed by deconstructing them into their constituent parts or underlying elements governed by fundamental laws or principles.

This concept is crucial in modelling physical systems for control systems analysis and design, and it is particularly dominant in the analytical mathematical modelling approach. In most modelling approaches, the underlying belief is that understanding the behaviour of the system under study can be achieved by studying the interactions and behaviours of its subsystems or components. The assumption is that the properties and behaviour of the system are determined solely by deriving the mathematical equations and relationships of its interactions and parts.

However, it can be argued that reductionism in many complex systems leads to oversimplifications, resulting in models that only partially and, under certain conditions, exactly describe the actual system behaviour. These simplifications also neglect emergent properties that arise from interactions between the parts and cannot be fully understood by studying those parts in isolation.

Finally, *mechanism* refers to a modelling approach that observes the physical system functioning like a machine or a complex system of interconnected mechanical parts. The mechanistic modelling view derives algebraic and dynamical equations based on deterministic and predictable *cause-and-effect* relationships. While this strategy works well for simple and certain more complicated systems with necessary assumptions, it *oversimplifies* the complexities of complex and adaptive systems, neglecting their emergent properties and holistic nature.

5.4.2 Ontology of Model-Based Analysis and Design

This section explores the *ontology* of model-based analysis and design approaches. From an ontological perspective, system dynamics modelling can be understood through three distinct approaches: *Realism*, *Idealism*, and *Moderation*.

In the realism approach, it is assumed that a model exists for every natural phenomenon, implying that reality can be fully represented through modelling. This perspective treats models as direct reflections of the physical world.

In contrast, the idealism approach views models not as complete representations of reality but as conceptual abstractions shaped by the designer's perspective. According to this perspective, models are intellectual interpretations of natural phenomena rather than exact replicas. Idealism acknowledges the limitations of models in capturing the full complexity of real-world systems and emphasises the role of human creativity and conceptualisation in the modelling process.

The moderate approach takes a middle ground, recognising that while models can effectively describe many systems, there are cases where phenomena are too complex to be mathematically modelled. This approach balances the utility of modelling with an awareness of its limitations, particularly when dealing with highly complex or adaptive systems [10]. The moderate approach recognises both the usefulness of models in capturing essential behaviours of systems and the inherent limitations imposed by the complexity of certain phenomena. It emphasises the need for cautious interpretation and application of models when the underlying systems are highly complex and poorly understood.

5.4.3 The Model Generation Process

It is interesting to see how the models are generated in a general context. The general modelling cycle is depicted in Fig. 5.1, where explanation and confirmation are central elements. Explanation forms the foundation of understanding physical phenomena, built on *assumptions*, *hypotheses*, *theories*, and *laws*. These assumptions and hypotheses may change in real-world applications, requiring corresponding updates to theories and laws. The explanation process relies on two key mechanisms: *idealisation* and *unification*. Idealisation simplifies complex empirical facts into concise, manageable statements. However, this simplification often involves omitting certain details, leading to *unmodeled dynamics*. For instance, when modelling a physical system, some aspects of its behaviour may be excluded to maintain simplicity, resulting in gaps between the model and reality.

Idealisation may involve isolating a phenomenon from its environment and other external elements akin to the principles of reductionism. This process often neglects unknown disturbances, noise, and interactions, which are natural consequences of idealisation in modelling. Additionally, unifying seemingly unrelated phenomena is another key aspect of the explanation process. Unification seeks to connect disparate

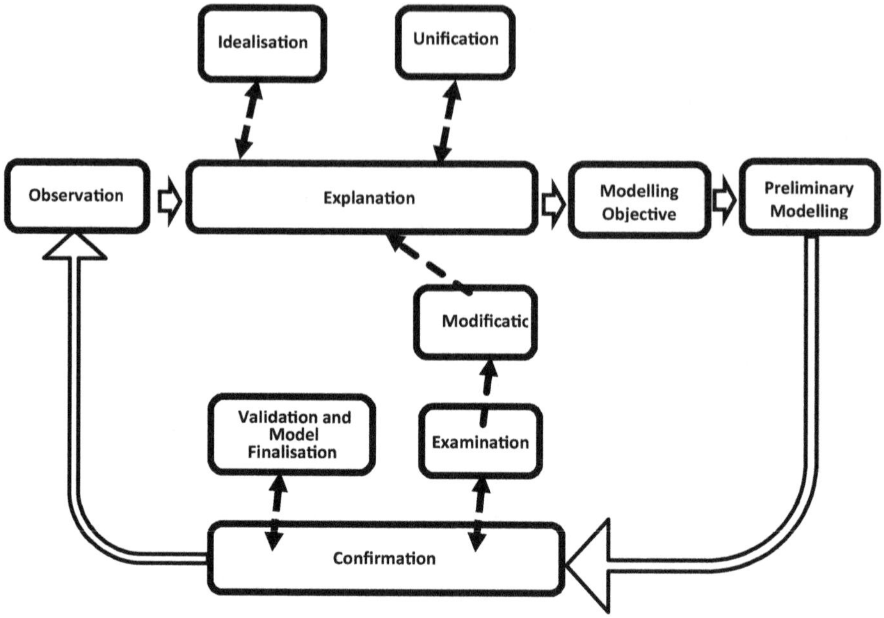

Fig. 5.1 The modelling cycle

observations or theories into a cohesive framework, enhancing our understanding of complex systems.

For example, consider the transfer function approach used to model different plants from various engineering disciplines. The transfer function approach is a common technique used to describe the input–output behaviour of dynamical systems. It is claimed that a transfer function of specific order and relative degree can effectively capture the dynamic response and behaviour of numerous systems in various engineering domains, given that different physical systems share similar underlying dynamics.

The transfer function approach provides a unifying framework to represent the behaviour of seemingly unrelated plants using a common rational function. By identifying the similarities and commonalities in the dynamics of different plants, engineers can create a unified transfer function model that captures the essential characteristics of those plants. For instance, even two different mechanical systems, such as a mass-spring-damper system, rotational system, or electrical circuit, can all have the same transfer function.

In the philosophy of science, *confirmation* refers to data and events that validate and support a scientific theory. This process involves conducting various tests and experiments to confirm the theory or law, thereby validating the underlying assumptions and hypotheses. Confirmations can be either qualitative or quantitative, depending on the nature of the evidence and the methods used. Model confirmation can be considered an overarching process encompassing the initial examination and subsequent validation steps in mathematical modelling and system identification.

Model examination, validation, and confirmation form a hierarchical and iterative process in model development. Examination serves as the foundational step, focusing on theoretical checks of a model's structure, assumptions, and parameters. Validation builds upon this by empirically testing the model's predictions against real-world data to ensure accuracy and generalisability. Together, these steps contribute to the broader process of confirmation, which encompasses validation and iterative refinements, such as adjusting the model based on early insights from examination or preliminary tests. Thus, while validation is a discrete, rigorous phase within confirmation, confirmation represents the end-to-end effort to certify a model's reliability, integrating theoretical scrutiny and empirical verification.

5.4.4 Some Remarks on the Inductive-Deductive Approach to Modelling

In Plato's view, understanding a constantly changing world is impossible [11]. He argued that the natural world is in a state of continuous change, making true cognition unattainable through sensory perception alone. Instead, Plato believed everything has a potentially perfect form,[3] and true knowledge lies in understanding these forms. He famously defined knowledge as *justified true belief* [12], and advocated for deductive reasoning as the path to acquiring it.

Deductive reasoning is a logical process that begins with a general idea and reaches a specific conclusion, often described as moving from the general to the specific. In scientific inquiry, this approach involves developing a hypothesis based on existing theory and then designing experiments to test it. This is sometimes called *top-down thinking* or moving from the general to the specific.

In contrast, Aristotle emphasised the importance of inductive reasoning in establishing basic assumptions before scientific investigation. Inductive reasoning involves drawing general conclusions from specific observations [13]. This process

[3] This refers to Plato's Theory of Forms (or Ideas), which is a central concept in his philosophy. It basically implies that beyond the physical world, there exists a realm of ideal or perfect Forms. Each tangible object in the physical world corresponds to a perfect Form that embodies its true essence. For example, the Form of a *circle* represents the perfect, unchanging concept of a circle, while physical circles (like those drawn on paper) are merely imperfect representations. The physical world is seen as flawed and transient. It consists of objects that change over time and can be perceived differently by different individuals. For example, a chair may look different under various lighting conditions or might be designed in multiple styles, but all these variations are just approximations of the perfect Form of *Chairness*. Also, while individual objects can strive to embody their Form, they never fully achieve perfection in the material world. This creates a distinction between the ideal (the Form) and the actual (the physical manifestation). Finally, Plato believed that true knowledge is acquired not through sensory experience (which can be misleading) but through intellectual reasoning that leads one to understand these ideal Forms. This pursuit involves recognizing the limitations of sensory perception and seeking deeper understanding through philosophical inquiry.

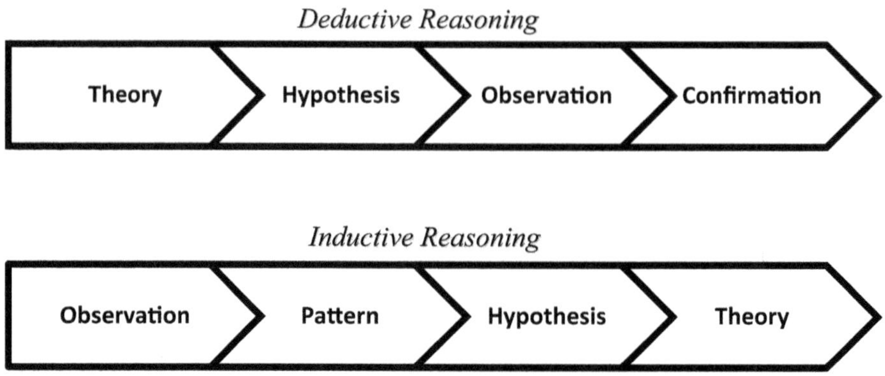

Fig. 5.2 Inductive reasoning versus deductive reasoning

starts with data, and conclusions are drawn from the data. Aristotle used inductive logic to establish causal links between *premises* and *hypotheses*. He introduced the concept of the *first principle* (*Primum movens*), which asserts that knowledge is built by gaining experience from what is already known to be true. Aristotle believed in the science of observation and measurement to derive general rules and construct models.

Aristotle viewed scientific research as a cyclical process: starting with observations, moving to general principles, and returning to observations for validation. He argued that scientists should deduce explanatory principles from observations and then use these principles to derive theorems about phenomena deductively. In Fig. 5.2, both inductive and deductive reasoning are briefly demonstrated.

To focus on the implications of the above argument in the context of modelling, it is argued that the system modeller uses inductive logic to develop a system model based on available information and data. Inductive reasoning enables the modeller to generalise from specific data, creating a model representing the system's behaviour. The model is then validated using the available information to confirm its accuracy.

Once the model is developed, deductive logic is applied to verify whether the model aligns with the system's known principles and available information. This dual approach, combining inductive reasoning for model creation and deductive reasoning for validation, ensures that the model is both data-based and theoretically sound.

However, the inductive-deductive approach to modelling dynamical systems presents several challenges, which are briefly summarised below.

- The model derived from the inductive-deductive approach can be *overfitted*[4] which makes it less applicable to different scenarios.

[4] In the context of modeling and machine learning, overfitting refers to a situation where a model learns not only the underlying patterns in the training data but also captures noise or random fluctuations that may not be representative of the true relationships in the data.When a model is overfitted, it performs exceptionally well on the data used for training but struggles to generalise to

- Inductive reasoning relies on the available information and data. If these are incomplete, biased, or incorrect at their time of application, the resulting model may incompletely or incorrectly represent the *true* system behaviour.
- Assumptions and simplifications result in models that can not fully capture all the *complexities* of the real system.
- *Validating* complex dynamical models is always a challenging issue, especially for systems with limited observability or during extreme operating conditions.
- Modelling complex dynamical systems requires developing accurate models with numerous variables, interactions, and operating points leading to *large* and *computationally intensive* models with little practical use.
- Models developed through inductive reasoning are designed to satisfactorily perform within the available limited information and data range in the modelling process. Such models may exhibit poor performance in unexpected and unpredictable situations. Model *robustness* to unforeseen conditions or variations in system behaviour is a critical point in practical applications.

5.5 Mathematical Models for Feedback Control

Mathematical modelling of dynamical systems revolutionised the analysis and design of control systems, shifting from trial-and-error heuristic methods driven by personal inspiration to the systematic processes of classical, modern, and postmodern control theories. Model-based control design methodologies have evolved through three distinct phases. Classical control relied on simplified linear input–output models, and modern control advanced this by introducing state-space models to handle more complex dynamics, yet both approaches ignored model uncertainty. This gap led to postmodern control, which explicitly recognised and addressed model uncertainty.

5.5.1 The Certainty Equivalence Principle

The linear input–output and state-space internal models are developed using the physical laws governing the system or system identification methods. Early researchers in the field often assumed that these models closely approximated the *true system*. However, this assumption proved inaccurate in many real-world applications.

The *certainty equivalence principle* is a fundamental concept supporting many model-based design techniques. This heuristic principle allows using approximate plant models for controller design, even when the estimation error is not negligible. As a result, the philosophy of model-based control can be summarised as [14]: *Estimate the best possible model, then design the controller based on this estimated model.*

new, unseen data. Essentially, the model becomes too tailored to the specific characteristics of the training data, making it less adaptable to different scenarios or datasets.

The Certainty Equivalence Principle was initially introduced in the context of optimal control theory. It is utilised to simplify the analysis and design of control systems under uncertainty. In practice, there are always unknown disturbances and uncertainties and accounting for all possible disturbances and uncertainties in the system can be complex, computationally demanding, and impractical in most cases.

However, the Certainty Equivalence Principle is introduced with a simplifying assumption to make model-based designs feasible. Under certain conditions, the controller designed for the system with all uncertainties and unknown disturbances ignored is applied to the true plant. In other words, a *key* assumption is that the controller designed based on the approximate system model is assumed to be effective in the true system.

For example, consider the Linear Quadratic Gaussian (LQG) control of a system with both state and measurement noise. The Certainty Equivalence Principle suggests that an LQG controller, designed for the deterministic model of the system with exact states and measurements (ignoring noise), can be directly applied to the true system. Remarkably, this approach yields satisfactory closed-loop performance even in the presence of noise.

Furthermore, in the context of adaptive control, the Certainty Equivalence Principle treats parameter estimates obtained from an identification algorithm, such as the recursive least squares in self-tuning regulators, as if they were the true system parameters in the controller. Consequently, despite the parameter estimates varying at each sampling time, they are effectively considered the true parameters by the Certainty Equivalence Principle [15].

Although the Certainty Equivalence Principle is considered a powerful and convenient tool to simplify the design and analysis of control systems in the presence of certain types of uncertainty and disturbance, its applicability can be seriously questioned in the face of complex systems. In such cases, robustness and adaptation issues should be brought forward to handle the complexities that may arise.

The mathematical proof of the Certainty Equivalence Principle requires strict and severe assumptions. It is assumed that the system can be modelled as a linear time-invariant dynamical system with *exactly known* system matrices and plant order subject to *known* state and measurement noise.

It can, therefore, be concluded that although the Certainty Equivalence Principle is a theoretically sound and beneficial concept in model-based design techniques, it is restricted when applied in practical systems for the following reasons:

- Practical control systems frequently defy precise deterministic modelling. Even minor noise characterisation or system parameter inaccuracies can cause substantial performance degradation, often leading to instability or poor closed-loop behaviour. Traditional design approaches rely on idealised models and noise assumptions, thus risking failure when deployed in real-world conditions where uncertainty and variability dominate.
- The Certainty Equivalence Principle ignores the errors that can occur due to sensor noise or other factors, which can impact the effectiveness of the control strategy.

- Actuator limitations are present in all practical systems neglected in the Certainty Equivalence Principle.

5.6 Model Uncertainty

The Certainty Equivalence Principle simplifies the control design process by ignoring discrepancies between the model and the actual system. Model-based control techniques often offer tractable and analytically solvable solutions, even for complex systems. However, their widespread application in real-world systems has been met with questions. In practice, the accuracy of the model used is often not practically acceptable, leading to significant differences from the true system. As a result, additional robustness measures are necessary to address *uncertainties* and improve performance in real applications.

It is stated in [16] that *there is no certainty, only less uncertainty*. This statement is particularly valid for control system design. In the philosophy of science, researchers have referred to uncertainties in engineering designs as *practical certainties* [16]. While founded on mathematical principles and validated through simulations under specific assumptions, control system designs often face impracticalities when implemented in real-world scenarios. Such designs may result in unstable closed-loop responses, poor performance, or performance degradation over time. These issues can often be attributed to the presence of uncertainties. To address these challenges, engineers must acknowledge and manage *uncertainties* during the design process, ensuring that control systems function reliably and effectively in practical applications.

5.6.1 Model Uncertainty Types

Uncertainties are broadly categorised into *aleatory* and *epistemic*. Aleatory uncertainties are random or stochastic in nature, while epistemic uncertainties arise from a lack of knowledge. Both types are often present in systems, and distinguishing between them can be challenging in practice [16]. To elaborate further on these issues, consider the following points:

- Aleatory uncertainty, also known as *random* uncertainty or *stochastic* uncertainty, arises due to the inherent randomness or stochastic nature of signals or systems. These uncertainties cannot be controlled or predicted with certainty and represent the natural variability in real-world systems.
- Epistemic uncertainty is related to limited understanding or lack of knowledge about a system. It can arise from simplifications made in system models for control system design. Unlike aleatory uncertainties, which are inherent and uncontrollable, epistemic uncertainties can be reduced or eliminated through further research, better data collection, or improved models.

Consider a practical example of epistemic uncertainty in an aircraft autopilot. The autopilot is designed to stabilise and control an aircraft using feedback from sensors, such as gyroscopes and accelerometers, to adjust its engine speed and control surfaces. In this case, epistemic uncertainty arises from incomplete knowledge or inaccurate modelling of the aircraft's complex aerodynamics and engine propeller behaviour. The complexity of airflow around the aircraft, particularly during manoeuvres, makes it challenging to develop a perfect aerodynamic model. Additionally, uncertainties may arise from simplified assumptions or limited wind tunnel testing, leading to epistemic uncertainty in the model's accuracy. The dynamics of the engines under various operating conditions may not be fully known, or they may vary with wear and tear, further contributing to epistemic uncertainty.

In real-world systems, it is not always straightforward to determine the exact nature of uncertainties. Some uncertainties may be obvious and easily identified as aleatory or epistemic. For instance, as previously stated, sensor noise is a clear example of aleatory uncertainty, while incomplete knowledge about a system's parameters represents epistemic uncertainty.

However, in many cases, uncertainties may have complex interactions and combinations of random and knowledge-related elements. This can make it challenging to categorise them as purely aleatory or epistemic precisely.

Also, many real-world systems exhibit a mix of aleatory and epistemic uncertainties. For example, in a control system for a chemical process, variations in raw material properties (aleatory) and incomplete knowledge about certain chemical reaction rates (epistemic) may contribute to overall uncertainty. Understanding and managing such combined uncertainties requires a sophisticated approach considering stochastic and knowledge-based elements.

A technique that can be utilised in this regard could be the *sensitivity analysis* approach. Refer to Sect. 4.2.1 for some historical details. Where changes in uncertain variables are studied to show their effect on the outputs of a system or model. The most influential sources of uncertainty can be identified by quantifying the sensitivity of system outputs to different uncertain parameters.

Another approach is *uncertainty quantification*, which assesses the range and impact of uncertainties on system behaviour and performance. This is done through probabilistic methods or statistical analysis and estimation of the likelihood of different outcomes based on uncertain inputs.

Finally, Monte Carlo simulation can be used for uncertainty analysis. It involves running simulations with random samples of uncertain inputs to obtain probabilistic distributions of system responses. This gives insights into the system's behaviour under different uncertain scenarios. It is important to conclude that verifying and validating the control system design through testing and experimentation is critical in dealing with uncertainties. Experimental and real-world tests can uncover potential sources of uncertainty.

5.6.2 Sources of Model Uncertainty

The sources of uncertainty in simple systems include *time*, *randomness*, *contingency*, and *human error* [16]. These sources are also present in complicated, complex, and complex adaptive systems, as discussed in the following sections.[5]

In simple systems, uncertainty can arise due to *variations over time*. If the system output is recorded as a time series, its analysis can reveal the patterns, trends, and variations over time. Factors like ageing or time-dependent changes in the system's parameters introduce uncertainty as time passes. Random events or fluctuations can also create uncertainty in simple systems. These events may not follow a specific pattern or be predictable, leading to uncertain outcomes.

Uncertainty can stem from unforeseen events or conditions not considered when designing the control system. Some common contingencies in a control system that may be ignored in the original design are introduced. These include sensor or actuator *failures*, *environmental changes* such as sudden changes in weather, road conditions, or traffic patterns in an autonomous vehicle, cybersecurity attacks in network connectivity, and unanticipated disturbances that are not explicitly accounted for in the control system design.

Finally, the involvement of *human operators* in simple systems introduces the possibility of errors, mistakes, or unpredictable actions, which can lead to uncertainty in system performance. It can also be argued that the model uncertainty stemming from using simplified models or approximations or selecting inappropriate model structures and assumptions are consequences of human decisions and judgments. Model uncertainty often arises from the necessity to simplify complex real-world phenomena, and the choices made by human designers in this process can introduce errors.

Interaction is another source of uncertainty in complicated systems. It can cause an uncertainty magnitude amplification in multivariable systems. These systems involve a larger number of interacting subsystems and processes. This interaction increases the complexity of the system. When interactions among these components are intricate and non-linear, uncertainty in any subsystems can be amplified, and the effects of uncertain factors can be propagated throughout the system. This can result in challenging system behaviours that are difficult to predict.

Consider the example of the multivariable control system of an oil refinery process. The process involves multiple interacting subsystems and processes to convert crude oil into various petroleum products. The process control system in an oil refinery is a multivariable system that controls various output variables, such as temperature, pressure, flow rates, and chemical composition at different stages of the refining process. In this case, uncertainties can arise due to variations in the quality and composition of the incoming crude oil feedstock. Additionally, factors like changes in ambient temperature, humidity, and other environmental conditions can affect the process parameters. The uncertainty in the quality and composition of the crude oil feedstock can have amplification effects on the entire refining process. The quality

[5] For a general systems classification refer to Sect. 1.4.

of crude oil can vary over time and from different sources, leading to fluctuations in the properties of the feedstock.

These fluctuations can impact the efficiency and performance of various refining processes. Also, the interactions between different processing units, such as distillation towers, catalytic converters, and fractionation columns, can result in non-linear effects due to changes in feedstock quality. Small variations in the properties of crude oil can have complex and non-linear effects on the process outputs. Also, the uncertainty magnitude amplification in the oil refinery's multivariable control system can result in reduced product quality, increased energy consumption, and fluctuations in the production rates of different products. The control system needs to be able to handle these uncertainties and ensure stable and efficient operation.

Finally, in complex and complex adaptive systems having all the above sources of uncertainty for simple and complicated systems, *emergence* from the interactions and interconnectedness among the subsystems is another important uncertainty source.

Complex systems include numerous interconnected subsystems that interact in non-trivial ways. In such systems, emergence is the strategic characteristic. Emergence occurs when the interactions and interdependencies among subsystems lead to behaviours that could not have been directly predicted or explained solely by analysing the subsystems.

Also, complex adaptive systems, a more intricate class of complex systems with time-varying characteristics, are capable of *self-organisation* and *adaptation.* These system interactions and feedback loops can lead to novel and unexpected behaviours, contributing to uncertainty.

Examples of uncertainty due to the emergence in a control system can include the following cases. Self-organising multi-agent systems emergence can occur when individual agents interact and self-organise to achieve a collective goal. The resulting behaviour of the overall system might exhibit emergent patterns that were not explicitly planned or predicted. Traffic flow in smart cities is an example of how the interactions between traffic lights, vehicle sensors, and adaptive traffic control algorithms can lead to emergent traffic patterns that optimise overall traffic flow but were not directly designed or predicted. Smart power grids with distributed control systems and present interactions among various power generation sources and consumers can lead to emergent stability or instability and might exhibit self-regulating properties due to the interactions between different components.

In control systems incorporating machine learning and adaptive algorithms, emergence can arise when the system learns and adapts to its environment. The system's behaviour might evolve based on interactions with the environment, leading to unexpected emergent properties. Finally, the *butterfly effect* in chaotic systems is a classic example of how minor variations in initial conditions can result in dramatically divergent outcomes. This inherent sensitivity to initial conditions raises uncertainty, rendering long-term predictions exceedingly difficult. It is also a defining characteristic of emergent behaviour.

5.6.3 Epistemological Aspects of Model Uncertainty

Epistemology of model uncertainty is a philosophical investigation of the nature, scope, and limitations of model uncertainty. It deals with general questions such as how model uncertainty is acquired and what exactly model uncertainty is. It is, therefore, more general compared to the sources of model uncertainty discussed above. Sources of model uncertainty are specific factors that contribute to uncertainty within individual models, whereas the epistemological aspect provides a broader framework for understanding and dealing with uncertainty in models across different contexts.

Epistemological aspects of model uncertainty deal with the philosophical and cognitive dimensions of understanding and reasoning about uncertainty in models. These aspects are general and not limited to any specific type of uncertainty or model, such as those previously discussed, but apply broadly to any situation where uncertainty exists.

Epistemological considerations are concerned with the nature of knowledge, belief, and the limitations of human cognition in the face of uncertainty. They are relevant across various fields and can be applied to different models and types of uncertainty.

To finalise this point, sources of model uncertainty refer to the specific factors or reasons that contribute to uncertainty within a particular model, and these sources can apply to various models. They are more *context-specific*.

For example, consider a distillation column in an industrial plant, where process models simulate and optimise the distillation column's performance. Sources of model uncertainty, in this case, are simplification and assumptions made to represent the complex thermodynamic and fluid dynamics processes involved. These simplifications could include neglecting certain minor interactions or approximating heat transfer effects. Another source is the distillation column's parameter estimation, where various estimated parameters are related to the distillation column's geometry, thermophysical properties of components, and column internals. Model structure and data quality are the other sources of model uncertainty in the distillation column. Now, while these sources of model uncertainty apply to various distillation column models, they can be context-specific based on the plan's specific operating conditions and objectives. A distillation column used in a petrochemical or food processing plant can have significantly different components and operating conditions. Also, modelling the performance of a distillation column under steady-state or dynamic and transient conditions varies. Hence, the context-specific nature of sources of model uncertainty in a distillation column emphasises the importance of tailoring the model to the specific plant and process conditions. On the other hand, epistemological aspects of model uncertainty pertain to the philosophical and cognitive dimensions of understanding and representing uncertainty in models.

Engineers are primarily interested in sources of uncertainty and using them to handle the effects of uncertainty in their design. However, a deeper understanding can be achieved by studying the epistemological aspects of model uncertainty to get the ability to interpret, communicate, and reason about uncertainty in models. That is, to interpret and understand uncertainty in model predictions, express model uncertainty to other parties and maybe non-engineers effectively.

For example, in control systems engineering, the epistemological aspect of model uncertainty revolves around how stakeholders interpret and reason about uncertainty in the models used to design and implement control systems. When control systems engineers interpret model uncertainty, they must consider the implications for the system's stability, performance, and robustness. They may use tools such as sensitivity analysis, uncertainty quantification, and robust control theory to assess the impact of uncertainty on the control system's behaviour and to design controllers that can effectively handle uncertainties. Stakeholders such as project managers, regulatory authorities, and end-users may have varying levels of understanding of control system models and their inherent uncertainties. Project managers may need to make decisions based on the projected performance of the control system, while regulatory authorities may require evidence of the system's reliability and safety in the presence of uncertainties. Effective communication of uncertainty in control system models is crucial for ensuring stakeholders make informed decisions. The epistemological aspect of model uncertainty in control systems engineering also involves addressing the cognitive biases and subjective interpretations that stakeholders may have regarding uncertainty. Engineers must handle these perspectives to build trust and confidence in the control system's capabilities. By exploring how control systems engineers and stakeholders interpret and reason about uncertainty in control system models, we can gain insights into the epistemological challenges of designing and implementing robust and reliable control systems.

5.7 Model Uncertainty Mitigation

Engineers frequently rely on heuristic methods to address and mitigate uncertainty in engineering designs. Heuristics are practical, experience-based approaches that provide simplified rules or guidelines to tackle complex problems. While these methods effectively reduce uncertainty, they are not without limitations. In fact, heuristics can introduce new uncertainties, as they often involve simplifications, assumptions, or approximations. In model-based control systems, uncertainties are typically addressed through mathematical modelling. Models are abstract representations of real-world processes. Although they are one of the primary ways to mitigate uncertainties from whatever source they arise, they are imperfect representations of reality, and using them introduces new uncertainties, referred to as *model uncertainty* [16].

Designers often collaborate closely with system model developers in a control engineering environment. To meet design requirements, models are sometimes simplified, even though they are fundamentally developed to represent objective empirical phenomena. This collaboration ensures that the model structure and objectives align with the needs of the control system design process. Once developed, models are validated against predefined criteria using data from real-world experiments conducted under various working conditions. However, these experiments are finite and may not cover all possible operating modes of a system. Additionally, limitations in human rationality, cognitive capacity, information availability, time, economic resources, and technological capabilities constrain the modelling process. As a result, the process concludes once a model is deemed sufficiently *accurate* for control system design purposes.

Model uncertainty management is a fundamental engineering concern, and in particular, control system designers should employ various techniques to manage and reduce model uncertainty and its effects. Sensitivity analysis, robustness analysis and model adaptation strategies are the key tools used to enhance the reliability of models in the face of uncertainties. In a model-based design environment, control system designers often face trade-offs between model accuracy and complexity. More complex models may better represent system dynamics but could introduce higher uncertainties due to increased parameterisation. Designers must make informed decisions regarding model complexity based on the specific control objectives and available resources.

Another notable issue is *confidence and uncertainty quantification*, where control system designers employ probabilistic and statistical methods to quantify the uncertainty associated with the models. Confidence intervals and uncertainty bounds measure the model's reliability and help control system designers make informed decisions. In the case of complex systems with inherent uncertainties, employing hybrid approaches that combine standard analytical models with data-driven techniques can overcome some uncertainty-derived problems.

Moreover, *risk assessment* methods can sometimes identify critical uncertainties that could significantly impact the system's performance or safety. This can provide critical information for the control systems designer to seek safer closed-loop performance.

5.7.1 The Trio of Uncertainty Combat: Robustness, Adaptation, Intelligence

It is quoted from Erich Fromm, a notable German social psychologist, psychoanalyst, and philosopher, that "*The quest for certainty blocks the search for meaning. Uncertainty is the very condition that impels man to unfold his powers.*" Embracing uncertainty and using tools like feedback can increase resilience, drive innovation, and reveal potential opportunities. This principle resonates not only in personal development but also in complex industrial processes. In personal contexts, individuals learn to handle uncertainties through self-reflection and feedback from their experiences. Similarly, in industrial applications, feedback control plays a crucial role in adapting to variability in operating conditions. The ultimate task of feedback control is to ensure closed-loop stability and acceptable performance in the face of plant uncertainties and disturbances.

However, achieving this stability and performance should consider practical constraints, such as resource limitations, costs, and operational capabilities. Practical control systems must design feedback mechanisms to be cost-effective while maintaining high performance.

A fundamental aspect of feedback control systems is the concept of *responsibility* and *interdependence* between the input and output of the system. Feedback control systems are designed to monitor and adjust the input behaviour based on the system's output behaviour. They take *responsibility* for adjusting the input to achieve desired output behaviour under different conditions and environments. If the output deviates from its desired value, the control system takes corrective action to bring it back on track. This responsibility is a key feature of feedback mechanisms. For *interdependence*, the input is influenced by the output, and the input influences the output. This mutual influence is at the heart of feedback control, allowing the system to self-regulate and respond to changes, disturbances and unknown conditions. The modification or regulation process is continuous and dynamic, allowing the system to maintain or reach a desired state.

These encompass the essence of feedback for controlling uncertain or unknown time-varying plants with practically tolerable cost. The control feedback theory that disregards the essential uncertainty concept and fails to acknowledge the true cost of feedback is of limited or no practical value from a control engineering perspective.

Three philosophical approaches to the feedback control of real plants, developed since the mid-twentieth century, involve the features of *robustness*, *adaptation*, and *intelligence*. The control design methodologies based on these concepts were briefly historically reviewed in Chap. 4.

5.7.2 *The Model-Assumption Dilemma*

A key point underlying all model-based control systems design philosophy is that real systems are approximated with models of different forms and complexities. Mathematical models are employed in the most basic designs, assuming no notable uncertainty.

In the robustness approach, exact models are accompanied by uncertainties to form a range of plant parameters and structures for modelling error compensation. In the adaptation approach, the discrepancy between reality and models is remedied by online identification or estimations, and the models are continuously updated to find the best fit. In the intelligence approach, models are derived from a conceptually different context. Soft computing techniques provide models of the intelligence framework. These would provide models that mimic the true plant to give the best artificial behavioural fit.

Furthermore, these models are accompanied by accurate a *priori assumptions* about their structure, bounds, input signals, and other factors. Nevertheless, in the online application of these models, the required online information, in cases of unknown plant encounters or changing conditions, is often unavailable and inaccessible to the control system, and any attempts to derive such online data often rely on additional a priori assumptions about the system.

This situation can be capsulised in the statement that "one must assume something to know something".[6] This philosophical concept forms the foundation of all model-based control system designs. It is a fundamental principle in problem-solving, reasoning, and knowledge acquisition. This strategy implies that assumptions or hypotheses are necessary to gain knowledge. When learning a new concept, we often assume it adheres to certain rules or patterns. We might also assume the correctness of our understanding until proven otherwise.

However, it is important to recognise that assumptions are not fail-safe, and we should remain open to revising them based on new evidence or information. In the control system design philosophy, this concept translates into making certain assumptions about the system we aim to control. For example, we might assume that the system is linear, time-invariant, and possesses specific properties that enable using particular control techniques. These assumptions are crucial for developing a model and establishing uncertainty bounds and structures of the system, which are then used to design a controller. However, assumptions are not always accurate, and revising them when new evidence or information becomes available becomes necessary. Nevertheless, such revisions are not always feasible during the controller implementation, which can potentially lead to adverse outcomes.

5.7.3 Uncertainty Combat: The Robustness Approach

From a philosophical point of view, the concept of *robustness* can entail discussions such as resilience and strength. It might refer to the resilience or strength of an argument, theory, or philosophical position that can withstand challenges, criticisms, and counterarguments, and it remains persuasive or valid even under various conditions. It may also refer to the comprehensiveness and versatility of a framework or theory that accounts for a wide range of phenomena, handles new information, and applies to various situations without requiring significant modifications.

In epistemology, robustness may be associated with the reliability and responsibility of knowledge or beliefs. A robust belief system or epistemological framework would be one that can justify and defend beliefs against scepticism and doubts.

The philosophy of robust control methodologies to handle uncertainty in real systems is rooted in designing control systems that can effectively operate despite various sources of uncertainty. Instead of designing good controllers for just one

[6] This is attributed to George E. P. Box, a famous statistician and engineer who pioneered the use of Bayesian methods in industrial applications. He also coined the phrase "all models are wrong, but some are useful", which reflects his view that models are always simplifications of reality, but they can still provide valuable insights and predictions.

model of the true plant, design a controller for a whole family of models that mimics the true plant under different conditions. In this case, the single-designed fixed controller, which is non-adaptive or non self-tuning, ensures closed-loop stability and desired performance for all models in the family set.

Robust control systems often involve worst-case analysis, where the controller is designed to perform well under the most adverse conditions within specified uncertainty bounds. This approach ensures that performance remains acceptable even when uncertainty is at its worst.

There is typically a *trade-off* between the robustness of a control system and its performance. Moreover, robust controllers may sacrifice some performance to ensure stability and acceptable operation under uncertain conditions. Feedback allows the controller output to adjust based on observed system behaviour in real-time, compensating for uncertainties as they arise. The other limitations of robust control are discussed in Sect. 4.2.3.

5.7.4 Uncertainty Combat: The Adaptation Approach

From a philosophical point of view, *adaptability* or *adaptive* may refer to the ability of theories, concepts, or frameworks to respond to new information, changing circumstances, or unexpected challenges. In epistemology, adaptability refers to the capacity of an epistemological framework or theory to incorporate new evidence or information and adjust its beliefs or knowledge accordingly. An adaptable epistemology acknowledges that knowledge can evolve with new insights and experiences.

Also, adaptability is important in the context of scientific theories. Scientific theories need to be adaptable to new experimental results and observations. An adaptable scientific theory is open to modification or revision when confronted with empirical data that challenges its predictions. Hence, adaptability can philosophically imply a willingness to adjust, refine, or reconsider previous positions, theories, or concepts in light of new information, challenges, or evolving circumstances.

The philosophical backbone of adaptive control is rooted in enabling a control system to learn and adapt to changing or uncertain environments and system dynamics. This would contradict the fixed structure and parameter approach of robust control design. Uncertainty modelling in the context of robust control is also revised.

The key philosophical principles associated with adaptive control can be summarised as follows:

- *Continuous Learning*: Adaptive control is founded on the principle of continuous learning.
- *Agility and Flexibility*: Adaptive control systems are designed to be agile and flexible. They can adjust their control strategies to accommodate variations in the system or environment.
- *Model-Free or Model-Based*: Adaptive control can be both model-free and model-based.

- *Feedback and Observations for Error Reduction*: Adaptive control relies heavily on feedback and real-time observations of the controlled system's behaviour.
- *Robustness to Uncertainty*: While robust control and adaptive control share similarities, adaptive control focuses explicitly on handling uncertainties by continuously adjusting control parameters or strategies.
- The theory underlying treating identified or estimated values as the actual values is the *certainty equivalence principle*; refer to Sect. 5.5.1 for more details [17].
- Most adaptive methodologies, as in robust control, depend on *prior assumptions* about the system. For a discussion on this topic, refer to Sect. 4.3.3.

5.7.5 Uncertainty Combat: The Intelligence Approach

Philosophically, intelligence is often linked to the notions of *agency* and *autonomy*, which refer to the ability to act independently and make decisions based on one's own free will. Intelligence enables the ability to make *rational* choices. Artificial intelligent (AI) systems attempt to replicate or mimic human cognitive processes, such as learning, problem-solving, reasoning, and decision-making.

Self-awareness in the AI context refers to the ability to reflect on internal states and processes. It is the capacity to identify the system as a distinct entity with a unique environmental perspective. In intelligent control systems, controller self-awareness is a crucial philosophical concept that plays a significant role. This concept is the cornerstone of a control system's ability to monitor its performance and take corrective actions. One specific aspect of self-awareness that must be considered is *objective self-awareness*. Objective self-awareness involves the control system's ability to objectively assess its performance, comparing it to a predefined standard or desired performance.

When an intelligent control system encounters a performance error or deviation from the desired outcome, it relies on its objective self-awareness to recognise the discrepancy. This self-awareness enables the system to identify when it is not operating optimally or aligned with its intended objectives. The acknowledgement of this performance error is similar to a philosophical notion of self-awareness in which the system recognises its *imperfection* in relation to its ideal state.

However, self-awareness alone is not the endpoint; it is the foundation for the system's corrective measures. In the spirit of autonomy and agency, borrowed from philosophical principles, the intelligent control system exploits its self-awareness to make rational decisions. It initiates adjustments to align its performance with the desired standards. This process of *self-awareness*, *error recognition*, and *autonomous correction* embodies the essence of agency and autonomy in the context of control systems, mirroring the philosophical ideals that humans use to act independently and make rational choices.

By integrating the philosophical concept of self-awareness into intelligent control systems, they are empowered to function more efficiently, adapt to changing conditions, and continuously improve their performance, ultimately aligning themselves with their intended goals and objectives.

In contemplating the nature of truth, one is compelled to acknowledge its inherent complexity and ambiguity. It seems truth is not a singular entity easily encapsulated by any one viewpoint in isolation. Instead, it indicates the inclusion of diverse perspectives, each contributing to a mosaic of understanding. Acknowledging diverse viewpoints contributing to a more comprehensive truth can be paralleled with self-awareness.

Assigning absolute validity to any particular viewpoint is a dubious endeavour. In this philosophical discourse, the essence of truth lies in its complicated nature, resisting reduction to a singular, unequivocal perspective.

This approach involves an appreciation for partial truths, that become apparent within specific contexts. In navigating the vast landscape of uncertainty, one must be accustomed to the delicacies of imprecision and ambiguity, which, paradoxically, contribute to the human capacity for rational decision-making amid distortions.

Herein lies the core of modern relativism, acknowledging that truth is not an absolute entity but rather a dynamic, context-dependent phenomenon. The recognition of this relativistic paradigm lays the groundwork for a methodology in the design of intelligent systems. Inspired by the complex workings of the human mind, which skillfully deals with uncertainty, these philosophical reflections hint at a deep connection between the principles of relativism, embracing partial truths, and the intellectual foundation shaping the development of intelligent systems. In this mix of philosophy, we discover the threads of understanding as we explore the complexities of truth in our quest for knowledge.

Soft computing methodologies and their hybrids are widely employed to augment AI and integrate human expert knowledge into computational processes. These applications encompass the creation of intelligent autonomous systems/controllers and managing complex systems with unknown parameters, such as forecasting the world economy, overseeing industrial processes, and predicting geological changes within Earth's ecosystems. Notably, these methodologies demonstrate proficiency in processing information, adapting to dynamic environmental conditions, and acquiring knowledge from their surroundings.

In contrast to analytical methods, soft computing methodologies emulate consciousness and cognition in several crucial aspects: they possess the capacity to learn from experience, generalise to domains lacking direct experience, and, utilising parallel computer architectures that replicate biological processes, expedite the mapping from inputs to outputs more rapidly than inherently serial analytical representations.

However, this efficiency comes at the cost of reduced *accuracy*. If a certain tolerance for imprecision can be accepted, it becomes conceivable to broaden the range of applications, even extending to problems where analytical and mathematical representations are readily available. The driving force behind such an expansion lies in the anticipated reduced computational load and the subsequent acceleration of computation speeds, enabling more robust control.

The trade-off in the *efficiency-accuracy dilemma* is addressed by *the principle of increasing precision with decreasing intelligence*, a foundational concept in hierarchically intelligent control. Intelligent controllers are designed to operate autonomously in uncertain environments, requiring new design approaches from engineers. These approaches often integrate concepts from artificial intelligence, soft computing, and control theory to enable anthropomorphic tasks with minimal operator intervention.

In [18], the principle is analytically formulated using a three-level hierarchical structure, organisation, coordination, and execution, with methods for implementing each level in intelligent controllers. The principle is expressed in terms of entropy, aligning with formulating the hierarchically intelligent control problem, where the solution minimises total entropy.

The philosophical backbone of intelligent control revolves around designing control systems that can make autonomous, informed decisions and adapt to various situations using artificial intelligence and advanced computational techniques. Here are some key philosophical principles associated with intelligent control:

- *Autonomy and Learning*: Intelligent control systems are designed to operate autonomously, make decisions, and take corrective actions. Machine learning and artificial intelligence techniques are often used to facilitate this learning and adaptation process.
- *Data-Based Decision-Making*: The philosophy of intelligent control is fundamentally data-based. These systems use data to understand the system's current state and make decisions based on this information.
- *Optimisation*: Intelligent control systems aim to optimise control strategies in real time, seeking to achieve the best possible outcomes while considering constraints and objectives.
- *Robustness to Uncertainty*: Intelligent control often incorporates uncertainty modelling and decision-making under uncertainty, using probabilistic and adaptive techniques to maintain stability and performance.
- *Cognitive Control*: The concept of cognitive control involves control systems that exhibit advanced cognitive abilities, such as perception, reasoning, and decision-making, enabling them to adapt and operate effectively in complex and uncertain environments.
- *Self-Optimisation*: Intelligent control systems aim to self-optimise their performance, striving for continuous improvement without external intervention.

As a final point, the adaptive and, in particular, robust control systems are fundamentally designed to respond to a control law formulated by the control systems designer. Even with a highly non-linear mathematical model, the controllers only respond successfully to the control systems' predicted scenarios. This design philosophy, at its core, allows for satisfactory performance and avoids implementation failures by constantly monitoring the system and making improvements under the guidance of the control systems designer. On the other hand, intelligent learning strategies, particularly the newly proposed deep learning methodologies, can provide appropriate responses in tasks where complex online decisions have to be made.

This idea is reinforced by the fact that complex industrial systems continuously generate and record a large amount of data for various control systems and essentially human-based system monitoring. These data can be effectively utilised to train deep neural networks. Also, feedback to the intelligent algorithm is available almost instantaneously from various points in the plant.

5.8 Conclusions

The groundbreaking notion of employing mathematical models to analyse and design control systems for physical processes marked a paradigm shift in control theory and engineering practice. These mathematical models are examined from a broad philosophical perspective. The modelling approaches are classified into two main types: general analytical mathematical and data-driven modelling approaches. The prevalence of models in the design of feedback control systems is analysed through a philosophical framework based on induction, reductionism, and mechanisms. Additionally, an ontological perspective on system dynamics modelling is explored, examining realism, idealism, and moderation as distinct approaches for comprehending the intricacies of such systems. The chapter outlines a general modelling cycle founded on the fundamental concepts of explanation and confirmation. It investigates the implications of inductive and deductive reasoning within the modelling context. It argues that the scope of mathematical modelling for dynamic systems encompasses classical, modern, and postmodern control design techniques. Subsequently, the certainty equivalence principle is introduced as a pivotal topic in model-based control systems design. The challenge of model uncertainty is addressed as a critical issue in applying models in control theory and practice. This encompasses various aspects, including types of model uncertainty, sources of uncertainty, epistemological considerations, and strategies for mitigating model uncertainty. Finally, the trio of strategies for combating uncertainty, namely, robustness, adaptation, and intelligence, is revisited philosophically.

References

1. F. Dörfler, Data-driven control: part two of two: hot take: why not go with models? IEEE Control Syst. Mag. **43**(6), 27–31 (2023)
2. L. Zadeh, On the identification problem. IRE Trans. Circuit Theory **3**(4), 277–281 (1956)
3. B.R. Gaines, System identification, approximation and complexity. Int. J. Gen. Syst. **3**(3), 145–174 (1977)
4. B.R. Gaines, General system identification—fundamentals and results, in *Applied General Systems Research: Recent Developments and Trends* (Springer, 1978), pp. 91–104
5. A. Khaki-Sedigh, *An Introduction to Data-Driven Control Systems* (John Wiley & Sons, 2023)

6. G. Pillonetto, A. Aravkin, D. Gedon, L. Ljung, A.H. Ribeiro, T.B. Schön, Deep networks for system identification: a survey. arXiv preprint arXiv:2301.12832 (2023)
7. M. Nouri-Manzar, A. Khaki-Sedigh, From inductive model-based control system design to deductive data-driven unfalsified adaptive control: a philosophical perspective. J. Control **18**(1), 81–89 (2023)
8. P.N. Johnson-Laird, A model theory of induction. Int. Stud. Philos. Sci. **8**(1), 5–29 (1994)
9. D. Lewin, Engineering philosophy—the third culture? Aircraft Eng. Aerosp. Technol. **53**(12), 22–29 (1981)
10. E. Pruyt, What is system dynamics? a paradigmatic inquiry, in *Proceedings of the 2006 Conference of the System Dynamics Society*, vol. 29 (System Dynamics Society Nijmegen, 2006)
11. A. Gregory, *Plato's Philosophy of Science* (Bloomsbury Publishing, 2015)
12. K. Lehrer, T. Paxson, Knowledge: undefeated justified true belief. J. Philos. **66**(8), 225–237 (1969)
13. E. Heit, Properties of inductive reasoning. Psychon. Bull. Rev. **7**, 569–592 (2000)
14. P. Albertos, A.S. Piqueras, *Iterative Identification and Control: Advances in Theory and Applications* (Springer Science & Business Media, 2012)
15. K.J. Åström, Theory and applications of adaptive control—a survey. Automatica **19**(5), 471–486 (1983)
16. D. Michelfelder, N. Doorn, *The Routledge Handbook of the Philosophy of Engineering* (Routledge-Taylor & Francis Group, 2021)
17. H.A. Simon, Dynamic programming under uncertainty with a quadratic criterion function, Econom. J. Econom. Soc. 74–81 (1956)
18. G.N. Saridis, Analytic formulation of the principle of increasing precision with decreasing intelligence for intelligent machines, in *Robot Control 1988 (Syroco'88)* (Elsevier, 1989), pp. 529–534

Chapter 6
System Theory and Its Impact on Control Systems Design

In the case of all things which have several parts and in which the totality is not, as it were, a mere heap, but the whole is something besides the parts, there is a cause.
—Aristotle 980a Metaphysics, Translated by W. D. Ross

6.1 Introduction

During the early twentieth century, the dominant philosophical theory in many scientific and engineering circles was a form of *atomism*, which suggested that reality is composed of independent, indivisible entities known as atoms. This perspective was particularly associated with early positivism and logical atomism, which were supported by many proponents such as the prominent philosopher Bertrand Russell. Russell's views and other logical positivists emphasised a *reductionist* approach to understanding the world, asserting that complex phenomena could be broken down into simple, atomic components. This atomistic perspective on engineering sciences during the early 20th century influenced how scientists and engineers approached problem-solving and understanding complex systems. This reductionist approach focused on breaking down complex systems into their basic components, which could be studied individually and more easily understood. In engineering design, this perspective facilitated the development of modular systems, and complex systems were analysed and controlled by breaking them down into smaller, manageable components. However, it is essential to note that while the atomistic approach has been beneficial in many ways, it also has limitations. Some complex systems may not be fully understood or accurately modelled by simply breaking them down into individual components. In such cases, a more *holistic* or *systems-based* approach might be necessary to fully comprehend the behaviour and interactions of the components within the system.

In the early decades of the twentieth century, mathematician and philosopher Alfred North Whitehead challenged the mechanistic worldview that dominated scientific thought and speculated that reality comprised discrete, independent entities

A. Khaki-Sedigh, *Control Systems: A Historical and Philosophical Perspective*,
https://doi.org/10.1007/978-981-96-7426-8_6

operating in isolation. Whitehead rejected this notion, arguing for a more interconnected and dynamic understanding of reality. He proposed an *event-based* or *process ontology*, which emphasised that the fundamental *stuff* of the universe is not static matter but rather interconnected events and processes. These events, he argued, are deeply interrelated, with each event influencing and being influenced by others. This new perspective laid the groundwork for what Whitehead termed the *philosophy of organisms*, which today is widely recognised as *process philosophy* [1].

Parallel to Whitehead's philosophical contributions, biologist Karl Ludwig von Bertalanffy was developing a complementary framework known as *General Systems Theory* (GST). Bertalanffy, often regarded as one of the founding figures of *systems thinking*, pursued to create an interdisciplinary approach to understanding complex systems. He argued that biological, social, or technological systems are best understood not by analysing their components in isolation but by examining their interactions and relationships. This perspective was particularly influential in developing first-order cybernetics. The epistemological and scientific foundations linking cybernetics and general systems theory are explored in greater detail in [2, 3].

The main general principle in GST is that the interrelationships and interactions between elements define a system as a unified entity [4]. It also introduced rigorous mathematical descriptions and models, which were instrumental in advancing the theoretical and practical understanding of control systems. These developments facilitated the analysis and design of control systems and paved the way for engineering, biology, and social sciences innovations.

The contributions of Whitehead and Bertalanffy marked a radical shift in the understanding of complexity, interconnectedness, and systems. Whitehead's process philosophy highlighted reality's dynamic and relational nature, while Bertalanffy's GST provided a methodological framework for studying interconnected systems. Though initially developed in different contexts, these ideas converged to shape a new scientific and philosophical landscape that continues to influence contemporary thought and practice. They proposed a common understanding and description of dynamically interconnected systems. Their main commonalities are their *holistic* perspective, emphasising the importance of looking at systems as a whole and recognising that the behaviour and properties of a system cannot be fully understood by merely analysing its parts or subsystems in isolation.

Also, they emphasise *processes* rather than *static objects*, with dynamic interactions characterising processes and interactions between their components. In GST, the concept of feedback loops highlights the *interaction* and *influence* of subsystems on each other. At the same time, Whitehead's philosophy emphasises the *interconnectedness* and *interdependence* of elements within systems through *prehension*, which refers to how entities in the universe are affected by and incorporate aspects of other entities. Both process philosophy and GST recognise the concept of *emergence*, where new properties or phenomena arise from the interactions among system components.

However, it is important to point out that Whitehead's philosophy is more metaphysical and addresses fundamental questions about reality, creativity, and the nature of experience. In contrast, GST is a scientific framework that aims to describe and analyse complex systems across various disciplines. This emphasises the potential collaboration between philosophical and scientific approaches in advancing our understanding of complex systems and their control.

GST can be considered a subfield within the broader field of *system theory*. Another significant subfield is *Mathematical System Theory* (MST). While GST emphasises the holistic and conceptual understanding of the system, MST is more specialised. It focuses on using mathematical tools and techniques to model, analyse, and control the behaviour of systems. MST employs mathematical concepts such as differential equations, linear algebra, control theory, and signal processing to describe the quantitative relationships within systems and is the foundation of model-based control system theory.

6.2 General System Theory and Its Implications

GST is a multidisciplinary theoretical framework that seeks to describe and analyse complex systems, regardless of their specific nature or discipline. Biologist Ludwig von Bertalanffy first introduced it in the 1930s as a response to the growing recognition that traditional reductionist approaches in science and other fields were insufficient to understand and explain the interactions and behaviours of complex systems.

To initiate a discussion on the characteristics of GST and its implications, it is informative to examine the *status quo* of general systems analysis before the development of process philosophy and GST. Indeed, the understanding and description of dynamically interconnected systems were severely limited and fragmented. The analysis of systems behaviour was not well-established as a separate domain of study in science and engineering. The primary engineering and scientific approach to systems analysis and design was based on reductionist methods. Each system was studied in isolation, and no *common features* were recognised among systems from different disciplines. This is also known as *disciplinary silos*, where there is isolation and a lack of communication or collaboration between different academic or professional disciplines, as well as various scientific and engineering fields working independently without sharing information, resources, or ideas. Also, *cause-and-effect* relationships were typically assumed to be simple and unidirectional in the designs, and the designers did not adequately and directly address feedback loops.

In GST, systems can be classified as *open* or *closed* based on their interactions with their environment. Open systems interact with their environment and exchange matter, energy, and information with the external environment. These interactions allow the systems to adapt and balance their surroundings dynamically. Open systems are characterised by their ability to receive input from the environment, process it internally, and produce output that influences the external environment. On the other hand, closed systems do not interact with their environment and do not exchange matter, energy, or information with external surroundings.

In a closed system, the interactions and processes are confined within the system boundaries, and the system is considered to be isolated from its environment. It is important to distinguish between the two terminologies in GST and control system theory. In control theory, an open-loop control system is a system with no feedback control, and the control action is determined in advance. Whereas, a closed-loop control system is a system with feedback that employs feedback from the output to adjust the control input and regulate the system's behaviour.

Feedback loops defined in GST are mechanisms that allow a system to self-regulate and adjust its behaviour based on the feedback it receives from its environment or its output. With the advent of GST, hierarchical organisations are readily derived, and several systems are organised hierarchically to form larger systems with a structure of increasing complexity. Undeniably, during the first three decades of the twentieth century, it can be argued that systematisation emerged as a central driving force across nearly all areas of human activity, profoundly shaping industry, commerce, and many other areas. Evidence supporting this assertion can be observed across a wide range of domains. The scientific management movement,[1] pioneered by figures like Frederick Taylor, sought to optimise efficiency through systematic approaches to work processes. Similarly, the time and motion studies conducted by Frank and Lillian Gilbreth[2] aimed to improve productivity. Establishing journals such as *Systems* further highlighted the growing emphasis on structured and interdisciplinary thinking. In industry, Henry Ford transformed manufacturing systems[3] with the principles of systematisation and standardisation. Meanwhile, Morris Leeds[4] contributed to this trend by developing data-driven methodologies to determine fair and equitable wage rates, reflecting a broader shift toward systematic analysis in labour economics. Beyond industry and commerce, modernist art, architecture, and literature movements also embodied this spirit of systematisation. Modernism

[1] A management philosophy and approach that emerged in the late 19th and early 20th centuries, notably associated with the work of Frederick W. Taylor, often referred to as *Taylorism*. This movement aimed to improve efficiency and productivity in industrial and manufacturing settings through the application of scientific principles and methods.

[2] Frank B. Gilbreth and Lillian M. Gilbreth, a husband-and-wife team of industrial engineers who made significant contributions to the field of efficiency and productivity improvement in the early 20th century. Time and motion studies involve the systematic observation and analysis of work processes and tasks to identify inefficiencies and optimise productivity. The Gilbreths were pioneers in this field, and their work aimed to reduce wasted time and effort in various industrial and manual labor settings.

[3] Henry Ford, the founder of the Ford Motor Company, in the early 20th century. Ford is widely recognised for revolutionising the automotive industry and, to some extent, modern manufacturing as a whole through his groundbreaking production methods, particularly the development of the assembly line.

[4] Morris L. Leeds was an American industrialist, engineer, and a key figure in the development of early management theories and practices. He played a significant role in the emergence of the scientific management movement, which aimed to increase efficiency and productivity in industries through systematic approaches and data-driven decision-making. Leeds also founded the Leeds and Northrup Company, a manufacturer of precision instruments and timekeeping equipment.

emerged as a response to the social, technological, and cultural transformations, emphasising order, abstraction, and a break from traditional forms [5].

Abstraction was a fundamental aspect of all forms of systematisation. This process involved condensing the most critical elements from a vast collection of details, enabling these abstracted essentials to serve as the foundation for comparison, extrapolation, and redesign. By focusing on the core characteristics of a system or phenomenon, abstraction allowed for identifying patterns, principles, and relationships that might otherwise remain obscured. Within the framework of science and technology, the process of abstraction inherently requires measurement. Sir William Thomson[5] appropriately stated in 1883, "when you can measure what you are speaking about and express it in numbers, you know something about it". Measuring instruments played a pivotal role in advancing the process of systematisation. In 1924, the Taylor Instrument Companies, in a series of advertisements titled *The Sixth Sense of Industry*, pointed out that these instruments could not only supplement but also replace our five senses to control production processes. They aimed to "take the guesswork out of manufacturing." These advertisements aligned with the principles of scientific management, reflecting Taylor's scepticism of ordinary workers' ability to self-regulate. Taylor believed that workers were inherently prone to inefficiency and required constant oversight to ensure productivity. This perspective insisted that employees must be meticulously monitored and guided to adhere to what Taylor termed "the one best way". The need for human judgment or discretion was effectively eliminated by embedding this principle into an automatic control system. This approach reinforced managerial control and highlighted the trend of replacing human decision-making with mechanised processes, reflecting the growing faith in technology and systematisation as tools for achieving efficiency and order [5].

6.2.1 A Philosophical Perspective of System Theory

An influential figure associated with systems philosophy is Ervin Laszlo, who contributed to the interdisciplinary approach to understanding the organisation and functioning of systems. His philosophical approach looks at the world as a set of interconnected and interdependent systems, emphasising relationships and patterns rather than isolated components.

Laszlo's philosophical approach to system theory is similar to that of Bertalanffy, who shared a common interest in understanding systems and their principles. This philosophical approach aims to integrate knowledge gained from studying the natural world and suggests a holistic approach that considers various disciplines and perspectives to understand complex systems comprehensively [6].

---

[5] Sir William Thomson, later known as Lord Kelvin, was a distinguished Scottish physicist, mathematician, and engineer. He is renowned for his contributions to thermodynamics, formulation of the Kelvin scale of absolute temperature, and pioneering research in electricity and magnetism.

Bertalanffy was aware of and approved Laszlo's efforts, and he provided the preface for Laszlo's book. Laszlo describes the structure of this philosophy as [7]: "Envisage the structure of science as an edifice built on a horizontal two-dimensional plane, and rising above it in the third dimension. The two-dimensional foundation is *nature* as it is presented to science in observation and experiment. The structure rising above it is the body of concepts, principles, theories and laws evolved in a given science. I shall call 'empirical adequacy' the factor which requires of science to make contact between the theoretical structure and nature at all points where such contact is possible (i.e. where the constructs have 'epistemic correlations', 'rules of correspondence', or 'operational definitions'.) Such coupling of theory and nature is to be as extensive and intensive as currently feasible, given the existing theoretical structure. That is, theory should eventually match all relevant observations with quantitative and predictive accuracy."

This articulates a metaphorical representation of the structure of science using a three-dimensional analogy. It represents the foundational level of science, often referred to as *nature*. It encompasses the raw data obtained through observation and experimentation. Essentially, it is the empirical basis upon which scientific inquiry is built. The third dimension symbolises the theoretical framework of science. This includes concepts, principles, theories, and laws developed based on the observations and experiments conducted on the two-dimensional plane. It implies an elevation beyond mere data collection to constructing abstract ideas that explain and predict natural phenomena.

The crucial requirement for science to establish a meaningful connection between the theoretical structure and the observed nature is emphasised by *empirical adequacy*. This connection is expected to occur wherever possible and relevant. In other words, scientific theories should correspond to and accurately describe empirical observations. This alignment involves using concepts like *epistemic correlations, rules of correspondence*, or *operational definitions* to ensure a robust and accurate connection between theory and empirical reality.

6.2.2 General System Theory and Control Systems Design

The principles and concepts introduced by GST have been explored in the context of control system design. Although it has brought some fresh perspectives to the field, its direct impact on the traditional analysis, design principles, and implementation of control systems might be subject to interpretation and further investigation. GST is not inherently a mathematical theory, but its conceptual foundations have influenced various disciplines, including MST and control systems theory:

- GST's emphasis on interconnectedness and holistic views of systems could have encouraged control systems scientists to consider more complex and interconnected mathematical models.

- The study of complex systems, which aligns with GST's ideas, inspired control systems scientists to develop new techniques for modelling complex systems encountered in various disciplines.
- GST's vision of cross-disciplinary and interdisciplinary subjects and its promotion of interdisciplinary research within different scientific, engineering, and social systems have provided control systems scientists with the courage and motivation to address such subjects.
- GST's emphasis and initiation on emergent properties could have encouraged control systems scientists to explore the mathematical modelling of emergent phenomena within systems.
- GST introduced the concept of hierarchy in the context of complex systems theory. The control systems scientists had possibly adopted the idea of modelling hierarchical control systems, where different control layers address various levels of system behaviour.

Before the development of GST and the introduction of transfer functions and input-output models or state variables and state-space representations, the dominant philosophical framework in control systems design was rooted in a mechanistic approach. The following key features characterised this approach:

- *Determinism*: The system's behaviour was assumed to be entirely predictable based on its past outputs, implying a strict cause-and-effect relationship.
- *Reductionism*: The system's behaviour under control was thought to be fully explainable by analysing the laws and principles governing its components.
- *Primary Reactivity*: Changes in the system's behaviour were attributed solely to external environmental influences, with little consideration for internal dynamics or emergent properties.

This mechanistic foundation led to highly specialised and tailored control designs for specific systems, as a universal or general design framework did not yet exist. In philosophical circles, however, anti-mechanistic ideas began to gain acceptance and influence toward the end of the 19th century. These ideas emphasised the concepts of *wholeness* and the *system* as a unified entity rather than a mere collection of parts. This shift has been interpreted in the history of modern science as the triumph of *function* and *relation* over the traditional categories of *substance* and *thing* [3]. These philosophical developments had deep implications for control theory, paving the way for more holistic and interconnected approaches to system modelling and design. This shift represented a change in perspective toward understanding the world in terms of processes, interactions, and relationships rather than merely focusing on discrete physical entities.

Through the functional or black-box approach, control theory enabled the study of a system without requiring detailed *nomothetic* knowledge of its internal interactions. Instead, this approach focused on analysing the relationships between inputs and outputs. Later, with the adoption of a general systems perspective, the internal dynamics of systems were abstractly modelled using state variables, providing a more comprehensive framework for understanding and controlling complex systems.

After the introduction of GST, the control systems designers started to acknowledge that the interdependencies between different parts of the system play a crucial role in determining its performance. It can be argued that the GST's recognition of the importance of feedback contributed to the widespread adoption of feedback control in various engineering applications. Additionally, to address uncertainties, some control system designers incorporated adaptive and self-tuning techniques, aligning with GST's emphasis on self-regulation and adaptation within systems.

The interdisciplinary nature of GST opened the way for control system designers to investigate applications, connections, and analogies between different domains of science. Studies in biology, cognitive science, economics, philosophy, and social sciences have led to transferring knowledge and methodologies from other fields to enhance control system design and problem-solving approaches. In parallel, the concept of hierarchical organisation within systems, recognised by GST, has been applied in many industrial systems due to their intricate nature and the need to manage various control loops operating at different levels.

6.2.3 General System Theory and Cybernetics

Cybernetics is a transdisciplinary field that studies systems' structure, behaviour, and regulation, including both natural and artificial systems. The term *cybernetics* comes from the Greek word *kybernetes*, which means *steersman* or *governor*, and mathematician Norbert Wiener coined it in the 1940s. Cybernetics is closely related to GST and shares several foundational concepts, as both fields aim to understand complex systems and their behaviour. Focusing on the study of control and communication in systems, cybernetics has provided valuable insights into the design of automated control systems, biological systems modelling, cognitive science, the study of human-computer interactions, machine learning algorithms, and artificial intelligence-based control approaches.

The key notion of cybernetics can be summarised as follows. Feedback is the central concept in cybernetics, where a measured system's output is used to modify its behaviour in the form of *positive* and *reinforcing* or *negative* and *stabilising* feedback.

Cybernetics is concerned with feedback mechanisms to maintain the system's stability and desired performance. In this regard, cybernetics emphasises the role of communication and information flow in systems. It explores how information is transmitted, processed, and used to control the system's behaviour. In the face of developing complex systems and complexities, cybernetics explores self-organising processes in systems, where complex structures and behaviours emerge from simple interactions among system components. Cybernetics has also tackled the issue of *homeostasis*, where a system is desired to maintain its internal stability despite external changes. This is correspondingly called *first-order cybernetics*. Nevertheless, there were reservations about this idea, as expressed in [8]. "By the 1960s, many feared that Cybernetics was vacuous, and realised that it is not because something is fuzzy that it is deep." There was indeed a perception that cybernetics lacked meaningful content, depth, or substance, and some had realised that mere vagueness or fuzziness in concepts does not necessarily indicate profound or substantial ideas.

On the other hand, *second-order cybernetics (SOC)*, also known as *cybernetics of cybernetics*, emerged later in the 1960s. Second-order cybernetics includes a so-called *observer* as an integral part of the system under study. It assumes that the observer's perspective and interactions with the system influence the understanding and behaviour of the observed system, leading to the relationship between the *observer* and the *observed system*. The division between first-order and second-order cybernetics reflects the evolving nature of the field and its deeper exploration into the relationships between observers and the systems they study.

As discussed in Chap. 5, it is inherent in the modelling philosophy that our simplified representations facilitate our understanding of systems. This means that our models ignore certain aspects of the system that are irrelevant to the purposes for which they are constructed. As a result, it is essential to distinguish between the properties of the system itself and those of its model, which are shaped by the developer's perspective and goals.

Engineers working with mechanical systems often deeply understand their internal structure and behaviour, which can lead them to treat their models as if they were the actual system. This is because they tend to focus on the system's objective properties rather than considering the role of the observer in shaping their understanding. In contrast, second-order cybernetics recognises that systems are not passive entities but active agents interacting with their observers and environments.

When working with organismic or social systems, second-order cybernetics acknowledges that these systems are agents in their own right, interacting with another agent, the observer. However, in reality, observer and observed cannot be separated, and the outcome of observations will depend on their interaction. This highlights the importance of considering the observer's role in shaping our understanding of the system. In this context, the observer is also a cybernetic system attempting to construct a model of another cybernetic system. This recursive nature of observation and understanding underscores the need for self-reflection and awareness of one's own biases and assumptions when working with complex systems.

The focus of these cyberneticians on epistemological, psychological, and social issues provided a valuable counterbalance to the reductionist mindset that emerged after significant advancements in science and engineering [9]. Figure 6.1[6] shows the essence of the second-order cybernetics. This describes a diagram from a 1976 conversation between Stewart Brand, Margaret Mead, Gregory Bateson,[7] and Norbert

[6] This Figure is adopted from English Wikipedia and is under the Creative Commons Attribution-Share Alike 3.0 Unported license, which was redrawn from Co Evolution Quarterly and Cybernetics Society Conference brochure CoEvolutionary Quarterly, June 1976, Issue no. 10, pp. 32–44. Côté, M. (1976). https://commons.wikimedia.org/wiki/File:SOCyberntics.png.

[7] Stewart Brand, Margaret Mead, and Gregory Bateson are all influential figures in the fields of anthropology, sociology, and cybernetics. They were all connected to the Macy Conferences, a series of interdisciplinary conferences held from 1946 to 1953 that brought together experts from various fields to explore the ideas and applications of cybernetics. Stewart Brand (1939–2023)

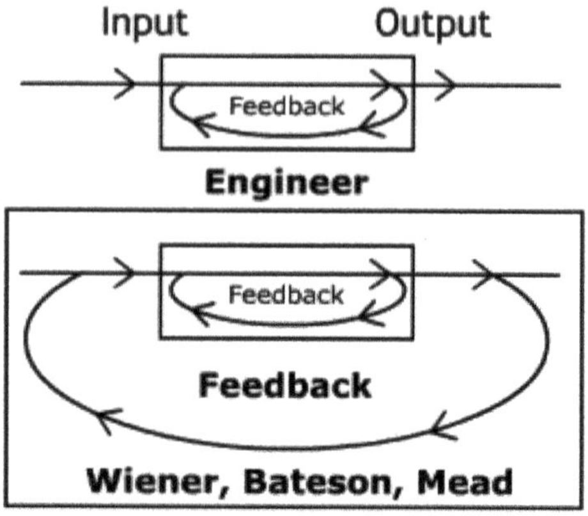

Fig. 6.1 First and second-order cybernetics

Wiener. In this figure, the input–output approach, which is a typical approach in engineering, a system is seen as a black box that receives inputs and produces outputs. Also, the observer is detached from the system and views it as a separate entity.

On the other hand, the participant observation approach indicates an active involvement in the system being observed. The observer is not detached but rather participates in the system, recognising that they are an integral part of it.

A concept similar to SOC, developed in the 1990s, is the notion of *Systems of Systems* (SoS). Both SoS and SOC have been developed to study complex systems and involve the consideration of observers. However, their focal points are different. SoS's primary task is the analysis of the integration and behaviour of interconnected systems in a specific application domain, whereas SOC deals with the epistemological implications of observation and the role of the observer in understanding and influencing the studied systems.

Another concept that is closely related to systems theory is the notion of a meta-system. A meta-system is a system that exists at a higher level of abstraction, observation, or regulation than the systems it affects. In other words, a meta-system is a system that observes, regulates, or interacts with other systems, often self-referential or self-modifying. This means that a meta-system can reflect on, modify, or even create its own rules, goals, or behaviour. In the context of the Principia Cybernetica Web,[8] a meta-system is exemplified by a child's playful behaviour. The child's

was an American author, journalist, and designer. Margaret Mead (1901–1978) was an American anthropologist and a leading figure in the field of anthropology. Gregory Bateson (1904–1980) was s British anthropologist, philosopher, and systems theorist and a pioneer in the development of cybernetics and systems theory. He was a key figure in the Macy Conferences and worked with Norbert Wiener. Bateson's work explored the relationships between human behavior, communication, and systems thinking.

[8] The Principia Cybernetica Web is a digital library and online encyclopedia of cybernetics, systems theory, and related fields. It is a comprehensive online resource that provides a vast collection of articles, documents, and resources on various topics related to cybernetics, systems theory, and complexity science. The website is designed to be a collaborative effort, allowing experts and scholars to contribute and share their knowledge and research.

actions, such as attempting to poke scissors into electric power points, creating ink patterns on the carpet, or drinking from a liquid detergent container, are driven by the intrinsic dynamics of the child's system, exploration and environment. However, this child's system is nested within a larger system, which is the childcare system. This meta-system operates according to its own autonomous principles and goals, aiming to provide a safe and nurturing environment for the child's development.

In systems theory, meta-systems are often used to describe complex systems that involve multiple levels of organisation, feedback loops, and recursive interactions. For example, a meta-system might be a set of rules or norms that govern the behaviour of individuals within a social group or a management system that regulates the performance of an organisation.

On the other hand, as previously mentioned, SOC focuses on self-referential systems, including meta-systems. This means that the system can reflect on its own behavior, modify its goals, and even observe its own observations. In other words, SOC deals with systems capable of self-reflection, self-awareness, and self-modification, leading to a more complex and dynamic understanding of system behavior. This framework also explains the emergence of new phenomena, such as autopoiesis[9] (self-creation and self-maintenance). Consequently, SOC provides a powerful framework for understanding the dynamics and behaviour of meta-systems, enabling the study of how these systems observe themselves, adapt their behaviour, and interact with their environment.

In all these cases, control systems design plays a crucial role in understanding and regulating the dynamics and behaviour of complex systems. Designing effective control systems requires a deep understanding of the system's nonlinear dynamics, self-referential properties, and interactions with other systems. In practice, cybernetics has complemented GST by focusing more on systems' control, regulation, and communication aspects. Meanwhile, GST has addressed the general principles and characteristics of complex systems. Over the subsequent decades, the two fields have collectively contributed to a more comprehensive understanding of how systems operate and interact, both in real-world applications and artificial intelligence.

---

[9] Autopoiesis is a term coined in the 1970s and refers to the self-organising and self-maintaining processes that occur within living systems, such as cells, organisms, and social systems. The term *autopoiesis* is derived from the Greek words *auto* meaning self and *poiesis* meaning creation or production. Autopoiesis can be understood as the process by which a system produces and maintains its own organisation and structure. Autopoiesis is characterised by Self-organisation, Self-maintenance, Autonomy, and Dissipative structures that maintain their organisation by dissipating energy and matter. The concept of autopoiesis has been influential in various fields, including biology, philosophy, sociology, and cognitive science, and engineering. Autopoietic systems in engineering can be seen in various applications such as self-organising robotic systems, and adaptive control systems. These systems are designed to mimic the self-organising and self-maintaining properties of living organisms, allowing for more efficient and resilient engineering solutions. By incorporating the principles of autopoiesis into engineering design, engineers are able to create systems that are more autonomous, adaptive, and robust, ultimately leading to more sustainable and innovative technologies.

6.2.4 Models and General System Theory

The development of mathematical modelling in the last century has profoundly shaped and advanced GST. By providing a rigorous and formal means to represent complex systems and their interactions, mathematical modelling has elevated GST principles to a more precise and quantifiable level, facilitating clearer communication and analysis. This formalisation has been particularly impactful with the emergence of systems dynamics, which shows how system variables respond to different inputs and serves as a platform for mathematically analysing the influence of feedback loops on system behaviour.

One of the key roles of mathematical modelling in GST has been to raise collaboration among researchers and engineers from diverse backgrounds and bridge the gap between disciplines by providing a common language and framework to address complex problems. Furthermore, optimisation techniques due to modelling complex systems have resulted in optimal solutions for system design and control.

Complexity theory, another product of mathematical modelling in GST, has significantly contributed to exploring emergent behaviours and self-organisation in complex systems. These phenomena can be better understood through mathematically based theories, offering new insights into complex systems.

Another benefit of mathematical modelling is its capacity to enable simulation studies in a controlled environment. This empowers engineers to test hypotheses, explore scenarios, and gain insights into system dynamics without real-world implementation. A practical example is mathematical modelling and simulation's estimation and predictive capabilities, allowing predictions about future system behaviour under different conditions and estimating variables that are not readily measurable.

Despite the substantial influence of mathematical modelling on GST, it is essential to acknowledge that not all aspects of GST require mathematical formalism. Many methodologies within GST remain qualitative and conceptual, emphasising systems thinking, emergent properties, and feedback mechanisms without explicit mathematical representations. The fruitful combination of mathematical modelling and qualitative approaches has enriched GST's versatility and applicability across various domains, making it a powerful framework for understanding and holistically addressing complex systems.

6.3 Mathematical System Theory

Although GST and MST are subfields of systems theory and share common characteristics, MST distinguishes itself from GST in several key ways. MST places greater emphasis on the use of mathematical models and formalisms for system analysis. It is firmly grounded in mathematical rigour, aiming to develop precise models that enable the mathematical analysis of system behaviour under various conditions. This focus allows MST to address critical challenges such as stabilisation and control. It is a fundamental tool in control systems, signal processing, and other engineering and scientific domains where mathematical modelling and analysis are essential for system design and optimisation.

6.3.1 Early Developments of Mathematical System Theory

The development of MST has involved contributions from numerous researchers over time, making it difficult to attribute its origins to a single individual. However, introducing mathematical models into control systems theory and design can be traced back to James Clerk Maxwell's seminal 1868 paper on governors [10]. In his paper, Maxwell proposed for the first time to draw the attention of engineers and mathematicians to the dynamical theory of governors. He did so without delving into any details of the mechanism. Maxwell's key contribution was the use of differential equations to model the behaviour of the flyball governor. This groundbreaking work marked the beginning of mathematical modelling in the stability analysis of closed-loop control systems, an idea that quickly gained attention and resolved many previously unsolved stability problems.

Although the differential equations describing the flyball governor were inherently nonlinear, Maxwell demonstrated remarkable ingenuity by linearising them. Through this process, he introduced concepts now recognised as real poles, imaginary poles, and the critical importance of pole placement in the right half-plane. Using differential equations, this model-based approach represented a pioneering milestone in the evolution of control theory. As such, Maxwell is often credited as a foundational figure in MST. For more historical details, refer to Sect. 2.4.2.

In the early twentieth century, classical control techniques emerged, laying the groundwork for analysing and designing control systems using mathematical methods. Pioneered by researchers such as Bode, Nyquist, Evans, and Nichols, these techniques relied on transfer functions, which could be derived from sets of algebraic and differential equations relating inputs and outputs. Alternatively, transfer functions could be obtained experimentally through simple tests on the plant, assuming linearity and time-invariance.

Later, in the 1960s, Rudolf Kalman introduced the state-space approach, which provided a more detailed and mathematically rigorous framework for modelling and analysing control systems. This approach marked a significant advancement in MST, enabling the study of more complex and dynamic systems. He is famous for developing various aspects of state-space theory, particularly the Kalman filter. These mathematical techniques were widely used in estimation and control systems, impacting various engineering and scientific fields. While numerous scientists and engineers, including figures like Richard Bellman, Lev Pontryagin, and Andrey Kolmogorov, collectively contributed to MST, Sanjoy Mitter and Allen Tannenbaum correctly assert in their work [11] that Kalman is the *father of mathematical system theory*.

In [12], Kalman assessed his predecessors' work in classical electrical network synthesis (which flourished between 1920 and 1970). The key points of his critique can be summarised as follows:

- *Over-Reliance on Intuition*: Kalman argued that early approaches to network synthesis were overly dependent on engineering and physical intuition. While valuable in some contexts, this reliance led to a lack of rigour in addressing the underlying mathematical challenges.
- *Avoidance of Mathematical Rigour*: Kalman criticised the tendency to sidestep mathematical issues whenever possible, describing it as "shoving the mathematical issues under the rug." According to Kalman, this avoidance hindered the development of a robust theoretical foundation for network synthesis.
- *Revival Through Mathematical Focus*: Kalman believed that the field could be brought back to life by concentrating on essential mathematics, much of which had been known since the 19th century. He envisioned a renewed focus on mathematical rigour as the key to unlocking spectacular prospects for scientific progress.

For more details on Kalman's legacy, refer to Sect. 3.3.2. However, Kalman's envisioned *Promised Land* encountered several challenging issues, which will be addressed and discussed in later sections.

6.3.2 State, Models, and Mathematical System Theory

Although the graphical approaches for analysing and designing control systems proved useful in many real applications, they did not provide an insightful picture of the system dynamics with rigorous mathematical foundations.

GST had no basis for the claim and, in fact, lacked the essential diversity and range of concepts, ideas, or components required to position itself as a *fundamental* interdisciplinary scientific endeavour, especially in the realm of control systems design. Also, Cybernetics, while generating numerous new ideas in control theory, did not constitute a comprehensive discipline for analysing and designing control systems. However, as addressed by Willems in [13], GST and Cybernetics paved the way and raised awareness that a theory for studying *man-made* systems and a discipline for addressing the challenges of *prescriptive science*, that is a discipline that addresses practical challenges and provides actionable solutions, were indeed lacking.

On the other hand, mathematics provides a precise language for expressing abstract notions and concepts and offers a structured framework for drawing logical conclusions based on these ideas. Neither GST nor Cybernetics fully embraced the idea that abstract concepts are best communicated and understood using mathematical language, as they were characterised by having a fundamental objective that was *abstract*. Their primary goals, aims, or purposes were rather conceptual and theoretical. They dealt with broad ideas, principles, and concepts that might not have direct physical manifestations but have implications and applications across various domains.

Kalman and other prominent researchers in control systems analysis and design filled this gap by combining a solid mathematical foundation with the ability and preference to envisage ideas that can have pragmatic solutions to control-oriented technological problems. The notion of *state* and using *state-space modelling* were pivotal components in this model-based control systems analysis and design methodology phase.

Other pioneering concepts within the context of MST include the *maximum principle* and *dynamic programming*. These are practical tools developed for solving optimisation and control problems, deriving optimal strategies that achieve desired performance while accounting for various constraints and objectives.

The state-space framework provided a mathematically concise and effective platform for modelling dynamical systems and performing estimation and control design tasks. This solid mathematical foundation yielded numerous results. Refer to Sect. 3.3.2 for more details.

The state-space-based system theory fundamentally transformed control system design thinking through the following key mechanisms:

- State-space representations developed a deep theoretical insight with a more comprehensive and intuitive understanding of the behaviour of dynamical systems.
- Adopting state-space models as a common representation for dynamical systems enabled control theory scientists and engineers from various disciplines to communicate effectively. This shared language facilitated discussions about analysis and design problems.
- The advent of state-space models led to the development of modern control concepts and techniques that revolutionised control system analysis and design. Rooted in mathematical rigour, the results ensured satisfactory closed-loop response across an expansive spectrum of applications. However, their effectiveness hinged on meeting the conditions and assumptions of relevant theorems and lemmas.
- State-space models introduced the capability to study the internal workings of dynamical systems, representing a departure from the traditional external input-output perspective in control system design.
- State-space representation naturally accommodated multivariable systems, allowing for the analysis and design of interconnected systems with multiple inputs and outputs.
- The emergence of modern control platforms provided control systems scientists and engineers with the unprecedented ability to explore previously inconceivable topics. This included engaging with issues such as estimation, prediction, and optimisation, which would have been considered beyond reach using frequency domain and s-plane transfer function models. The Kalman filter is a notable example of a remarkably influential concept within the MST. As one of

the most extensively employed tools, the ubiquitous Kalman filter has numerous applications spanning engineering, physics, computer science, and economics.

As highlighted in the previous point, the advent of modern control theory, with its focus on internal state modelling, paved the way for numerous new ideas in control theory. Two notable examples include *Lyapunov-based stability theory* and *optimal control design.*

Lyapunov-based Stability Theory. The concept of stability dates back to antiquity, with figures like Aristotle and Archimedes exploring related ideas. However, it was not until the Age of Reason in the seventeenth century that a more concrete, mathematical, and systematic approach to stability analysis in dynamical systems emerged.

A significant milestone occurred in the early twentieth century with the rise of Lyapunov's groundbreaking stability methodology in Western scientific discourse. This marked the beginning of modern stability theory, as highlighted in [14]. Lyapunov's work, originally presented in [15], gained attention in the English-speaking world and naturally aligned with the state-space representation and mathematical systems theory of modern control. As noted in [14], "Unfortunately, though his work was applied and continued in Russia, the time was not ripe in the West for his elegant theory, and it remained unknown there until its French translation in 1907. Its importance was finally recognized in the 1960s with the emergence of control theory."

In the history of control theory, Lyapunov's stability theory quickly became a cornerstone, establishing itself as the predominant tool for stability analysis within the control community. For more details on Lyapunov's legacy, refer to Sect. 2.5.

Optimal Control. The calculus of variations, with roots extending more than three centuries ago, can be recognised as the originator of optimal control theory, as elaborated in Sect. 2.4.1. While the calculus of variations has a longstanding history, optimal control theory, which focuses on determining the optimal control strategies for dynamic systems, emerged as a distinct field in the mid-20th century. The progress in MST and state-space analysis during the first half of the twentieth century played a pivotal role in making *optimality* feasible within modern control.

As suggested by the historical paper on optimal control [16], it is widely accepted that optimal control theory originated in the former Soviet Union. The primary basis for this assertion is the development of the Pontryagin maximum principle, which serves as a cornerstone of optimality theory. However, some mathematicians argue that this theory represents a modest extension of the classical calculus of variations, particularly by incorporating inequality constraints and control variables. According to this viewpoint, many challenges addressed by optimal control theory might have been resolvable within the framework of pre-modern control methodologies.

This perspective is supported by the reportedly limited attention the Pontryagin maximum principle received when it was announced at the 1958 International Congress of Mathematicians. It is asserted in [16] that non-mathematical factors might have been the primary cause of this negative reaction. The two main contextual factors cited for this claim are Pontryagin's personality, particularly his notorious anti-Semitism, and the perception that the result was primarily intended for military applications.

Optimal control theory constitutes a fascinating bridge between theoretical mathematical concepts and practical engineering applications. At its core, this theory revolves around the notion of determining the best control inputs to a dynamic system to achieve a desired performance while satisfying certain constraints. It provides a systematic framework for decision-making in scenarios where performance optimisation is crucial, such as guiding rockets, managing industrial processes, or controlling autonomous vehicles.

The calculus of variations, an essential mathematical tool, reinforces the foundation of optimal control theory by finding functions that extremise certain quantities, often formulated as *functionals*. In the context of optimal control, these functionals represent *performance indices* or *cost functions* that should be optimised. The fundamental idea is to find the trajectory or control inputs that minimise or maximise a given functional. However, translating theoretical mathematics to real-world engineering applications requires a perspective shift provided by MST.

In optimal control design, the focus extends beyond the abstract space of all possible curves considered in the calculus of variations. Instead, engineers are concerned with finding control strategies that adhere to the real constraints of the physical system, such as limitations in control inputs, dynamic behaviour, and practical considerations. Moreover, optimal control theory uniquely addresses minimum time problems that frequently arise in control engineering. These problems involve finding the quickest path for a system to transition from one state to another, subject to its physical limitations.

6.4 Limitations in Mathematical Systems Theory

Before studying the specific limitations and critiques of MST, it is valuable to address certain philosophical concerns that can be intertwined with its approach and applications. These are summarised as follows:

- By its nature, MST often employs *reductionist* approaches to break down complicated and complex systems into simpler mathematical models. Hence, the philosophical debates between *reductionism* and *holism* will arise in MST.
- Mathematical models used in MST often assume *deterministic* behaviour and *predictability*. Hence, the philosophical debates about *determinism versus indeterminism* in real-world systems arise.
- MST relies on mathematical reasoning and proof to analyse systems. This leads to questions about the nature of knowledge, certainty, and the validity of *mathematical modelling* in understanding reality.

MST, as a theoretical framework, has its own set of limitations. While it is a powerful tool for modelling and analysing dynamic systems, it is important to recognise its challenges in ensuring that its applications in practical real-world problems are feasible. Some of these challenges are the required assumptions and simplifications, causality and feedback, uncertainty and transparency, model characteristics being taken as system characteristics, and the theory and practice gap, which are discussed as follows.

6.4.1 Assumptions and Simplifications

Mathematical models of systems frequently rely on *assumptions* that may not align with real-world scenarios. Precise mathematical equalities and inequalities must be satisfied to apply the theoretical outcomes of theorems and associated algorithms. However, these assumptions can often oversimplify the complexities of practical systems, resulting in erroneous predictions and analyses and introducing uncertainties into the models. These uncertainties and inaccuracies can subsequently yield disparities between the model and reality, consequently impacting the performance of closed-loop control systems.

Numerous mathematical models within systems theory operate under the linearity assumption, implying a proportional connection between inputs and outputs. Nonetheless, real-world practical systems frequently have nonlinear behaviours. Also, the various linearisation techniques introduce extra inaccuracies and uncertainties into the models.

Furthermore, *model order reduction* techniques play a pivotal role within MST methodologies. When dealing with high-dimensional systems, the computational demands can become substantial, often necessitating simplifications or approximations.

Consequently, order reduction becomes indispensable for numerous analysis and design methods when implementing controllers in real-world applications. This introduces an additional layer of inaccuracies and uncertainties to the models.

Also, creating accurate mathematical models becomes increasingly challenging as systems become more complex. The complex nature of real-world systems might mandate the utilisation of simplified models, inevitably giving rise to inaccuracies and uncertainties.

Most mathematical system models presume a state of time-invariant behaviour, indicating that the system's characteristics do not change over time. In reality, systems can exhibit time-varying behaviour due to changing conditions, and ageing processes. Attempting to fit time-invariant models to such time-varying behaviours introduces a new layer of uncertainties.

6.4.2 Causality and Feedback

The connection between *cause* and *effect* lies at the heart of comprehending dynamic systems' behaviour and designing effective control methods. The philosophical doctrine that supports the relationship between cause and effect is known as *causation*. Control theory deals with managing the behaviour of systems to achieve desired outputs. This is achieved by manipulating system inputs or parameters based on their current and desired states. Hence, in control theory, causality refers to how changes in the input of a system (the cause) lead to changes in its output (the effect). This cause-and-effect relationship is crucial for understanding how the system responds to different manipulated inputs, disturbances, and noise.

Conversely, feedback loops play a significant role in control theory. A feedback loop occurs when a system's output is used as input to influence its subsequent behaviour. Control theory utilises these concepts to develop control algorithms and strategies that ensure the desired system's behaviour. By understanding the cause-and-effect relationships within a system and implementing appropriate feedback mechanisms, control theory helps achieve closed-loop stability and desired closed-loop performance. However, establishing true causality can be challenging due to various factors like *confounding variables*, *observational biases*, and *hidden interactions*. These factors are briefly introduced below.

Confounding variables are defined as those that are not of primary interest in control system design but can affect the relationship between the input-state-output variables and lead to false correlations. For example, consider studying temperature control systems in air conditioning systems. The primary goal of the temperature control system is to maintain a desired room temperature by adjusting the heating or cooling elements. It would be desired to understand the relationship between the room temperature, the heating/cooling input, and the output (temperature inside the room). However, other factors can influence the room temperature, such as external temperature changes, the number of people in the room, and the amount of sunlight entering through windows. These additional factors can act as confounding variables.

Observational bias is measurement errors, and hidden interactions are relationships that might not be immediately apparent and could involve indirect pathways or interactions not modelled in the mathematical models of the system.

Causality and feedback loops are not always properly understood and characterised in the idealised mathematical modelling of real-world systems. This deficiency can result in predictions and analyses that are prone to errors. Formulating conclusions grounded in *flawed causal assumptions* can, in turn, yield control system strategies that are practically ineffective.

It is important to note that while MST has these limitations, it remains a valuable tool for analysing and understanding many complex systems. However, users should be aware of these limitations and exercise caution when applying the theory to real-world situations.

6.4.3 Uncertainty and Transparency

While MST has proven invaluable in numerous theoretical and scientific inquiries, its shortcomings in practical engineering design applications have also been documented

[17, 18]. These malfunctions can be directly linked to the limitations discussed in Sect. 3.3.5. Subsequent examples from notable control theory scientists and engineers will be provided to illustrate these points.

In the classical paper [18], the authors explore the use of optimal control with a quadratic performance index as a method for regulating an industrial plant around a steady state. However, they argue and demonstrate that such control is often unattainable and, even when achieved, can exhibit significant drawbacks. The primary factor attributed to these limitations is *uncertainty*. Reference [19] expresses this as: "Lyapunov stability theory gained prominence in the early 1960s, yet it largely overlooked uncertainty. Similarly, optimal control dominated the 1960s, exhibiting a similar disregard for uncertainty."

The issue of *transparency* in control systems design can be considerably significant. Transparency, defined as the state of being transparent, allowing light to pass through, and characterised by clarity and diaphanousness, according to the Oxford Dictionary, holds a central place in global governance, where it is consistently sought after.

Transparency aims to achieve two key outcomes: *visibility* and *controllability*. When power (or aspects of system behaviour) is visible, its potential for effective control is enhanced. The concept of transparency has been addressed and claimed through two distinct approaches in control theory.

In the first approach, in an engineering approach to transparency, the advocates of robust control theory and professionals within the *quantitative feedback theory* (QFT) community have consistently argued that MST control system designs *lack* transparency [20]. It is claimed that the non-transparent nature of these methodologies exacerbates the *gap between theoretical concepts and their practical application* in scientific and engineering contexts. This theory-practice gap will be discussed further in this section. Transparency in the QFT philosophy is defined as visually correlating design parameters with real-world issues, starting from the initial design phase and continuing throughout each controller design stage [20].

In contrast, from a theoretical perspective, the second approach to transparency issues is formed in control design methods based on input-output or black-box models. These models provide limited insight into a system's internal structure. State-space representation, however, offers an internal description of dynamical systems. By selecting state variables that align with the physical system's parameters, designers gain an accessible view of internal interconnections and system behaviour.

Through concepts like *controllability* and *observability* (or *reachability*), along with associated assessment tests, designers can determine the extent to which inputs influence various system components and how control actions affect overall system performance. *Observability*, in particular, measures the effectiveness of system outputs in reflecting internal system variables, thereby enhancing system *visibility*. These concepts—controllability and observability—infuse system representation with a level of transparency that is absent in input-output approaches.

The insights derived from internal descriptions of dynamical systems have profoundly impacted control practitioners, solidifying the importance of state-space representation within the control community. For many control scientists, this shift not only marked the decline of classical control paradigms but also ushered in a new era of control design, underpinned by sophisticated mathematical frameworks. This transition was believed to effectively signalled the end of the pre-modern control theory era, paving the way for modern approaches.

6.4.4 System Characteristics Versus Model Characteristics

MST introduced numerous interesting and theoretically valuable concepts in control theory. These concepts include, but are not limited to, *controllability*, *observability*, *duality*, *minimality*, *realisations*, and *decoupling*. MST introduced these concepts alongside mathematical tests and algorithms to validate and compute them, under certain given exact model assumptions and conditions.

For instance, *controllability* is defined in terms of a specific state-space realisation of the system. A system described by a set of state-space equations is considered *state-controllable* if its corresponding *controllability matrix*, as defined, possesses full rank. Furthermore, a given transfer function can yield an infinite number of state-space realisations, which may or may not be controllable. A *minimal realisation* is subsequently defined through a specific set of equivalent conditions.

Willem's perspective on the concept of controllability offers valuable insights [21]. He argues that one of the key lessons from the development of system theory is that a minimal state-space system can represent every transfer function and convolution system. In this context, minimality is equivalent to controllability and observability. However, this has led to the impression that controllability is not an intrinsic property of a dynamical system itself but rather a property of its state-space representation, a specific way of modelling the system.

Despite this, the notion that a lack of controllability is linked to the presence of common factors in a transfer function (and can thus be viewed, in some sense, as an external property) remains deeply rooted in system theory folklore. As Willem points out, there are strong system-theoretic reasons to question this interpretation [21].

The concept of *decoupling*, along with its corresponding tests and procedures, is similarly defined with a distinct set of assumptions and conditions. Subsequently, a procedure is provided that effectively achieves exact decoupling for a multivariable system. Exact decoupling is theoretically appealing and gives an ideal closed-loop multivariable system. In [22], the authors assert that for a coupled multivariable *system* satisfying specific strict assumptions, the ability to be decoupled through state variable feedback, the class of all feedback matrices capable of decoupling the system can be fully characterised. In the context of the MST literature, the term *system* refers to a specific state-space representation with certain properties. Decoupling indices are defined, and using these indices, a decoupling matrix is constructed. This decoupling matrix must be of full rank for full decoupling of a multivariable system. Although the authors conclude by noting that practical implementation of the techniques discussed in their paper has already commenced, they emphasise that significant work remains before the theory can be fully developed into a practical design methodology. While the theoretical decoupling controllers were intellectually compelling, they did not evolve into practical design control systems. The later concept of *diagonal dominance*, introduced by Rosenbrock, aimed to address the practical limitations of the decoupling concept.

Hence, one can argue that these conditions and other similar conditions can be regarded as conditions at the *representation level*, as elucidated in [23]. In other words, they are characteristics of the specific state-space model adopted for the system and do not necessarily reflect the attributes of the *actual* system under consideration. Notably, the system state-space matrices constitute a representation realised by the user for the system.

Willems further advances the concept of the *behavioural system-theoretic framework* to define controllability in terms of the system's behaviour or trajectories. To elaborate, *past trajectories* can be seen as trajectories determined by nature, *future trajectories* as desired trajectories set by the designer, and the *controlled trajectory* judiciously chosen by the control engineer [24]. This comprehensive definition of *behavioural controllability* is rooted in system behaviour and holds relevance for general dynamical systems. In the case of LTI systems, it can be linked to the definition of conventional state-space controllability. However, it is important to note that the two definitions are not equivalent.

In summary, it is generally understood that the concepts arising from MST represent theoretical properties of mathematical representations of dynamical systems and do not inherently reflect genuine properties of the actual dynamical systems.

6.4.5 Theory-Practice Gap

During the early 1970s, the field of modern control theory began encountering unexpected issues. These issues were linked to failures in implementing the theories developed up to that point. Engineers actively applying modern control theory to practically design control systems faced significant challenges during this period. The theoretical principles that appeared sound on paper proved difficult to implement successfully in real-world applications. The root cause of the engineer's difficulties was model *simplifications* and *uncertainty*. The struggles and failures experienced by engineers triggered a sense of crisis within the control engineering community.

These challenges were not just academic or theoretical but were causing concern among professionals in the field. The failures were unexpected and had a significant impact on the way control systems were designed and implemented. As a result of these challenges, the disparity between the theoretical concepts and practical implementation in control engineering became evident. This *gap* between what was understood and proved in theory and what was achievable in practice became known as the *Theory-Practice Gap*[10] in the control engineering community. One of the earliest papers discussing this issue is 1963 [25], which states, "AT PRESENT there is a well-known gap between control theory and practice."

Later, in 1973, [26] tackled the problem from a process control perspective. The abstract of the paper states "During the past 10 years many of the techniques which precipitated modern control have been disseminated, digested, and applied in various fields of engineering. Some experiences with modern control have been favorable, others have not been. To a large extent, we have learned something about those types of problems for which modern control can be expected to do things that classical control is unable to; and, just as importantly, we have also learned that modern control cannot be used successfully in all applications." Directly addressing the chemical process control gap, it states, "Indeed, the theory of chemical process control has some rugged terrain to traverse before it meets the needs of those who would apply it." The theory-practice gap in control systems analysis and design has been tackled by many researchers, as shown in [27] and the references therein.

Before the advent of state-space mathematical system theory, practical control engineers frequently incorporated a substantial portion of control theory directly into real-world applications. The emerging gap highlighted the contrast between the idealised principles of control theory and the challenges encountered when attempting to implement them into practice within realistic, real-world scenarios.

In [28], the shift from practical control designs in the early twentieth century to the mathematically driven modern control approaches is described as a *reversal of progress*. The authors argue that the balance between theory and practice, achieved over decades, was disrupted, this time in favour of theory. As a result, the pure theory gained significant attention, leading to a perception among control engineers and scientists that a gap had emerged and the holistic view of control systems had been lost. Reflecting on the evolution of control theory, a prominent system theorist notes in [8] that the significance of that era only became clear in hindsight. At the time, the developments did not seem particularly extraordinary. He adds that one might have expected professional organisations and senior academics to embrace the advancements enthusiastically, but this was far from the case, and all the developed theory was a game for the young. The increasing mathematisation of the field was met with resistance; for instance, matrix exponentials were dismissed as abstract mathematics, while Laplace transforms were widely accepted as intuitive tools as everybody knew what they were. Despite the exceptional richness of ideas introduced through solutions to the linear-quadratic-Gaussian problem and the proliferation of system-theoretic concepts, these contributions failed to gain widespread acceptance. The fact that state-space systems could model a far broader range of phenomena than transfer functions was largely overlooked. A significant challenge at the time was the lack of concrete industrial applications, which hindered broader adoption.

[10] The phrase "theory-practice gap" is commonly used in various fields, including engineering, education, nursing, psychology, sociology, and organisational studies. It suggests that there is a disconnect between what is learned in theory and what is applied in practice. Although the exact origin of this phrase is unclear, but it has been in use since at least the 1960s.

Moreover, the motivation behind much of the work was often rooted in mathematics for its own sake rather than technological advancement [8].

To address this gap, control system scientists in the mid-1970s began developing new tools to tackle uncertainty within the MST framework. The initial response to the limitations of modern control theories in dealing with uncertainty was the pursuit of solutions within the established modern paradigm by control theory modernists. This endeavour is encapsulated in the words of one prominent control theory researcher whose biography states that his research interests and dedication are to make modern control theoretical results applicable in practice [29]. This line of research soon turned into new schools of taught in postmodern control theory, as discussed in Chap. 4.

Quantitative Feedback Theory (QFT) is a frequency response design technique that claims to bridge the theory-practice gap in control system design [30]. The founder of this approach argues in [19] that Ph.D. graduates from the most prestigious universities were often narrowly focused on achieving specific objectives through feedback. Those trained in classical control theory aimed for desired input–output relationships, while *modernists* prioritised eigenvalue realisation, an even more restrictive goal. This mindset, he contends, has led to the unnecessary labelling of any feedback theory addressing uncertainty as *robust*. In a practical design philosophy, uncertainty is not an afterthought but an integral part of feedback design, rendering additional labels like robustness redundant. It would be akin to specifying *live human beings* whenever discussing humans, an obvious and unnecessary distinction [19].

Analysing the theory-practice gap in control, it is argued in [27] that one of the causes of this gap is the fact that expertise in control systems analysis and design requires a strong foundation in abstract theory, particularly in applied mathematics. While control theory is typically taught in engineering departments, research in the field often involves a significant amount of mathematical analysis; hence, control theory is closely intertwined with mathematics in research and development. It is also observed that while control theory has relevance across various industries and fields, the emphasis on education and expertise often leans towards theoretical aspects rather than specific real-world applications. This is due to time constraints and the need to deal with students from diverse engineering backgrounds, which often leads control courses to use simplified examples, such as inverted pendulums and stirred tank reactors. These examples may not fully capture the complexities of control implementations in practical industrial environments like chemical plants or automobiles. Achieving truly innovative solutions requires a blend of in-depth domain expertise in the specific application and proficiency in control theory. The importance of collaboration and teamwork is underscored as essential for bridging the gap between theoretical understanding and practical implementation in control systems.

6.5 Conclusions

Process philosophy and systems theory are recognised as the two primary driving forces behind model-based control approaches. They offer a shared understanding and description of dynamically interconnected systems from various perspectives. The two subfields of systems theory are introduced as GST and MST. GST emphasises a holistic and conceptual understanding of systems, while MST is more specialised, focusing on the use of mathematical tools and techniques for modelling, analysing, and controlling system behaviour.

GST and its implications are studied, including the indirect impact on control system design. Furthermore, the connections between GST and cybernetics are highlighted. The role of models in GST is then addressed, illustrating the significant implications of modelling within the context of GST.

MST is examined more extensively than the other subfields of systems theory. It is argued that MST has profoundly impacted control systems, signal processing, and various engineering and scientific domains. The historical development of MST is discussed, focusing on the novel and groundbreaking state-space platform initiated and developed by Kalman. It is argued that mathematics provides a precise language for expressing abstract notions and concepts and offers a structured framework for drawing logical conclusions based on these ideas. Indeed, MST provides a solid mathematical foundation with the ability and inclination to think creatively and envision ideas that can provide pragmatic solutions to control-oriented technological problems. Although MST has proven to be a solid foundation for control systems analysis and design, its limitations are also presented and discussed.

References

1. A.N. Whitehead, *Science and the Modern World* (Macmillan, 1925)
2. M. Drack, D. Pouvreau, On the history of Ludwig von Bertalanffy's "General Systemology", and on its relationship to cybernetics–part III: convergences and divergences. Int. J. General Syst. **44**(5), 523–571 (2015)
3. D. Pouvreau, M. Drack, On the history of Ludwig von Bertalanffy's "General Systemology", and on its relationship to cybernetics: part I: elements on the origins and genesis of Ludwig von Bertalanffy's "General Systemology." Int. J. General Syst. **36**(3), 281–337 (2007)
4. L. Von Bertalanffy, General system theory, a new approach to unity of science. 5. Conclusion. Human Biol. **23**(4), 337–345 (1951)
5. S. Bennett, The past of PID controllers. Annual Rev. Control **25**, 43–53 (2001)
6. K. Adams, P. Hester, J. Bradley, A historical perspective of systems theory, in IIE Annual Conference and Expo 2013. 4102–4109 (2013)
7. E. Laszlo, Introduction to systems theory: Toward a new paradigm of contemporary thought, *New York: Harper Torchbooks and (1973) The Rise of General Theories in Contemporary Science*. J. Gener. Philos. Sci. **4**, 335–344 (1972)
8. J.C. Willems, In control, almost from the beginning until the day after tomorrow. Euro. J. Control **13**(1), 71 (2007)
9. F. Heylighen, C. Joslyn, Cybernetics and second-order cybernetics. Encycl. Phys. Sci. Technol. **4**, 155–170 (2001)

10. J.C. Maxwell, On governors. Proc. R. Soc. London **16**, 270–283 (1868)
11. S. Mitter, A. Tannenbaum, Rudolf E. Kalman: the father of mathematical systems theory [Historical Perspectives]. IEEE Control Syst. Mag. **37**(2), 164–166 (2017)
12. R. Kalman, Old and new directions of research in system theory, in *Perspectives in Mathematical System Theory, Control, and Signal Processing: a Festschrift in Honor of Yutaka Yamamoto on the Occasion of his 60th Birthday* (Springer, 2010), pp. 3–13
13. J. Willems, Dynamical systems, controllability, and observability: a post-modern point of view, in *Mathematical System Theory: the Influence of RE Kalman* (Springer, 1991), pp. 17–37
14. R.I. Leine, The historical development of classical stability concepts: lagrange, poisson and Lyapunov stability. Nonlinear Dyn. **59**, 173–182 (2010)
15. A.M. Lyapunov, The general problem of the stability of motion. Int. J. Control **55**(3), 531–534 (1992)
16. H.J. Sussmann, J.C. Willems, 300 years of optimal control: from the brachystochrone to the maximum principle. IEEE Control Syst. Mag. **17**(3), 32–44 (1997)
17. M.G. Safonov, Origins of robust control: early history and future speculations. Annual Rev. Control **36**(2), 173–181 (2012)
18. H. Rosenbrock, P. McMorran, Good, bad, or optimal? IEEE Trans. Autom. Control **16**(6), 552–554 (1971)
19. I.M. Horowitz, *Quantitative Feedback Design Theory (QFT), Volume 1* (QFT Publications, 1993)
20. C. Houpis, S. Rasmussen, M. Garcia-Sanz, *Quantitative Feedback Theory: fundamentals and Applications* (CRC Taylor & Francis, 2006)
21. J.C. Willems, Controllability and observability in a new perspective, in *16th Symposium on Operations Research* University of Trier 1991 (Springer Science & Business Media)
22. W. Wolovich, P. Falb, Decoupling in the design and synthesis of multivariable control systems (1967)
23. I. Markovsky, J.C. Willems, S. Van Huffel, B. De Moor, *Exact and Approximate Modeling of Linear Systems: a Behavioral Approach* (SIAM, 2006)
24. J.C. Willems, Paradigms and puzzles in the theory of dynamical systems. IEEE Trans. Autom. Control **36**(3), 259–294 (1991)
25. A. Fuller, Directions of research in control. Automatica **1**(4), 289–296 (1963)
26. A. Foss, Critique of chemical process control theory. IEEE Trans. Autom. Control **18**(6), 642–652 (1973)
27. T. Samad et al., Industry engagement with control research: perspective and messages. Annual Rev. Control **49**, 1–14 (2020)
28. K.J. Åström, P.R. Kumar, Control: a perspective. Automatica **50**(1), 3–43 (2014)
29. C. Harvey, G. Stein, Quadratic weights for asymptotic regulator properties. IEEE Trans. Autom. Control **23**(3), 378–387 (1978)
30. C.H. Houpis, S.J. Rasmussen, M. Garcia-Sanz, *Quantitative Feedback Theory: fundamentals and Applications* (CRC Press, 1999)

Chapter 7
Falsification Philosophy and Control Systems Design

The only real mistake is the one from which we learn nothing.
—Henry Ford

7.1 Introduction

Falsification is a concept that Karl Popper introduced in his work, which was first published in 1934 in German and later in English in 1959 [1]. *Falsificationism*, a theory within the philosophy of science, addresses the problem of *demarcation*. Demarcation pertains to distinguishing between scientific and non-scientific claims. It involves clarifying the boundary between statements or theories that are scientific, pseudo-scientific, or unscientific.

Popper's philosophy of falsification is a notable example of a primarily philosophical concept that has indirectly influenced engineering and applied sciences. It can be generally stated that the falsification philosophy is aligned with engineering practices in the following directions:

- Falsificationist principles align with the rigorous iterative testing and validation processes engineers use to ensure the reliability and performance of the designed systems. This is carried out to challenge the proposed solutions and identify potential *failures*. Failures can provide insights into design flaws, assumption limitations, uncertainties and other challenges.
- It is a common engineering practice to seek and predict possible failure modes and to enhance the designed systems to withstand a variety of conditions and scenarios, making them more reliable and *robust*.
- The falsification philosophy inspires engineers to identify and assess potential risks associated with their designs. In doing so, the *worst-case scenarios* are considered to falsify their design's assumptions.
- Applying falsificationist principles leads the engineer to better *quality assurance* practices, as the designer will actively seek weaknesses and potential points of failure.

A. Khaki-Sedigh, *Control Systems: A Historical and Philosophical Perspective*,
https://doi.org/10.1007/978-981-96-7426-8_7

The principles of falsification philosophy have influenced various aspects of control systems design, including system identification and modelling, controller design, performance testing, fault tolerance and safe control. By adopting a falsificationist mindset, designers can develop more robust and reliable control strategies that adapt to uncertainties and unexpected challenges. A notable example is the *unfalsified adaptive control*. These principles have provided a solid framework for emerging *data-driven control* methodologies [2].

7.2 Engineering Knowledge and Falsification: Adapting Philosophical Frameworks and Responding to Failed Designs

Engineering and science operate within distinct problem settings. Their activities and knowledge exhibit similarities and distinctions, which provide key comparative analysis points for our study. These include [3]:

1. *Similarities in Activities*
    - They involve developing and examining descriptive theories about the physical world.
    - Both disciplines engage in designing and constructing artefacts as part of their work.
2. *Distinctions in Objectives*
    - Understanding natural phenomena is the ultimate objective of science.
    - Creating practical solutions and artefacts or transforming the world is the ultimate objective of engineering.
3. *Similarities and Distinctions in Knowledge*
    - Both fields rely on empirical theories to describe phenomena.
    - Engineering knowledge extends beyond the behaviour and design of artefacts, including specifications of usage requirements and assertions that artefacts meet specified requirements.
4. *Methods of Cultivating Knowledge*
    - They share similar methods for generating knowledge, such as experimentation and observation.
    - Science has a more theoretical and exploratory approach, while engineering has distinct methods focusing on practical application and problem-solving.

For details on engineering ontology and a concise comparative study of engineering and science methods, refer to Sect. 1.3. Despite the differences and with appropriate modifications, the study of engineering knowledge, or *engineering epistemology*, can benefit from examining and adapting philosophical frameworks developed to understand scientific knowledge. By applying these existing philosophical frameworks to engineering, we can better understand how engineers generate, evaluate, and apply knowledge in their field.

Some key areas where philosophical frameworks from the philosophy of science can be adapted to engineering epistemology include the following topics:

- The nature of engineering theories and models.
- The role of experimentation and simulation.
- The relationship between theory and practice.
- The social and cultural aspects of engineering knowledge.

Engineering theories are utilised to show that artefacts meet the design requirements and specifications. These theories are expected to predict whether the artefacts will fulfil these requirements under defined conditions. However, they can be subjected to *falsification*, and in this context within engineering, falsification occurs when artefacts fail to fulfil their mission during real-world implementations. In such cases, an engineering *failure* has occurred.

To be more specific, in control engineering, *requirements* refer to the specifications and objectives a control system must achieve to meet the desired performance of a particular plant. These requirements can be broadly categorised into the following three main aspects. *Performance requirements* that specify the desired behaviour of the controlled system, such as stability, accuracy, speed, precision, and robustness. *Constraints* and *limitations* that impose restrictions on the control system's design and operation. *Operational requirements* that relate to the control system's integration with the overall plant intended to control.

Also, in control engineering, *artefacts* primarily refer to the physical or software components that make up a control system. These can include sensors, actuators, controllers, and other hardware or software elements that work together to monitor, control, and regulate the behaviour of a controlled system. Control system design algorithms are the key artefacts in control engineering. These algorithms are essential components of control systems, as they define the strategies and methods used by the system to achieve the desired performance and behaviour.

Identifying the potential reasons for engineering failures and the possible ways to respond to them is of prime importance. Multiple factors can contribute to the discrepancy between the designed artefacts and their intended usage. These factors include:

- *Invalid engineering theories*: The theories that connect designs to requirement specifications may contain flaws or inaccuracies that lead to incorrect predictions about the performance of the artefacts.
- *Inconsistencies in requirements specifications*: The specifications for the artefacts might not accurately represent the actual requirements for their practical applications, resulting in mismatches between the designed artefacts and their intended purpose.

Let us examine these concepts in a control systems framework. Invalid control systems engineering theories can arise from several sources, such as incorrect assumptions, where theories are based on incorrect or oversimplified assumptions about the controlled plant, which may not hold in real-world scenarios. Inaccurate modelling that leads to inadequate predictions about the performance of the control system. Insufficient data that can result in designs that do not adequately represent the system's behaviour.

Also, inconsistencies in requirements specifications defining a control system's desired performance, constraints, and operational aspects can lead to discrepancies between the designed control systems and their application in the real world. Some reasons for inconsistencies in requirements specifications include ambiguity of the requirements specifications, changing requirements that evolve over time due to changes in the controlled plant, and incomplete understanding in defining the requirements specifications, leading to inaccurate or incomplete specifications.

When engineering failures occur, there are three primary ways to respond to the falsification:

- *Changing requirements* (or *requirements specifications*): This involves re-evaluating and possibly modifying the specifications to better align with the actual requirements for the usage situations.
- *Changing artefacts* (or *designs*): This approach focuses on redesigning or modifying the artefacts to meet the specified requirements better.
- *Changing engineering theories*: If the underlying theories connecting designs to requirements specifications are found to be flawed, revising or improving these theories can lead to more accurate predictions and performance improvements.

In the case of engineering failures, engineers may resort to ad hoc adjustments in theories or utilise a falsification strategy to systematically challenge assumptions against empirical evidence. The ad hoc adjustments often occur without a thorough understanding of the underlying issues. This approach may raise concerns about the reliability and validity of engineering theories and their influence on the design and development of artefacts. On the other hand, utilising falsification in engineering, guided by the principles of *critical rationalism*, suggests that engineering theories should be subjected to rigorous scrutiny and falsification. This means that theories should be tested against real-world observations and data, and if they fail to explain or predict these observations accurately, they should be revised or even discarded. The focus is on continuously improving our understanding of theories by learning from failures and inconsistencies rather than relying on ad hoc adjustments or the *rescue* of theories when they face falsification.

Addressing falsifications in problem-solving can occur at different levels of generality, ranging from broad research contexts to more specific design contexts. In both cases, the goal is to identify and resolve issues related to falsifying theories or assumptions. In research settings, problem-solving involves testing and challenging broad theories or hypotheses. This may include investigating fundamental principles or concepts in a particular field of study. In these contexts, the focus is on understanding the underlying issues and refining theories to improve their reliability and validity. In design contexts, the problem-solving process is more focused and practical. It involves addressing specific issues related to the development and analysis of artefacts. Here, the challenge is ensuring that the theories and assumptions guiding the design process are accurate and effective in solving the problem. This level of problem-solving is crucial for successfully creating and implementing practical solutions.

Responding to falsification can lead to developing a completely new theory and approach. When a current theory or hypothesis is falsified or challenged, it often necessitates re-evaluating existing knowledge and understanding. This re-evaluation may create new ideas, theories, or approaches to explain the observed phenomena better or address the initial problem more effectively.

In some cases, these new theories or approaches may significantly impact their respective fields, leading to paradigm shifts. Such breakthroughs often stem from addressing falsifications and the subsequent need to adapt and innovate in response to the challenges they present.

Engineering theories are explicit, falsifiable claims that predict and analyse the performance of artefacts while considering their requirements. They critically incorporate practice-based empirical assertions, which are often limited in scope but essential for practical application with approximate claims. However, to be valid logically, the empirical theory must be true under all observable interpretations. Empirical theories often include constraints that define their limitations in scope or precision without rendering the theories inherently false. While such approximate theories may fall short of the rigorous standards required for scientific purposes, they can still be sufficiently accurate and practical for many engineering applications [3].

Engineering failures can be due to various sources. Two key contributors are (1) the inadequacy of theories in identifying design flaws that led to their application and (2) the insufficiency of necessary requirements, which may fail to align with real-world needs and environmental assumptions accurately.

The concluding remarks from [3] are as follows: Reject or change the analytical engineering theory if an artefact fails to meet its requirements despite the predictions of an engineering theory, but the corresponding design and requirements specification are valid, then the theory is invalid and must be rejected. Future theoretical analysis must use another theory. This may be an entirely new engineering theory, an alternative pre-existing engineering theory, or a revised version of the rejected theory. The rejection of engineering theories and the development of new ones is how engineering knowledge grows. Revising a rejected engineering theory will sometimes occur by weakening the rejected theory, particularly by approximation (reducing claimed precision or accuracy), or by qualification (adding extra conditions).

The condition for rejection/change of an engineering theory is if an artefact fails to meet its requirements despite the theoretical predictions and the design and requirements specification being valid, indicating no implementation errors. Upon the invalidity of a theory, it must be rejected or revised in one of the following ways: a completely new engineering theory should be developed, an alternative pre-existing theory should be used, or a revised version of the rejected theory should be adopted [3].

The knowledge creation process in research and design contexts is closely connected to control systems engineering. Research findings can influence design approaches, while the design process itself can lead to new insights and theories. The control system designs can range from highly specific solutions tailored for a particular system to highly parametric designs that can be adapted to address a broader range of problems. As control systems designers work on designing and analysing specific control problems, they may find new design principles that can be applied to a wider range of problems.

7.3 The Falsification Theory

7.3.1 A Scientific Theory

Popper argues that no finite number of observations can verify general theoretical claims, rejecting inductive proof. He also asserts that *observational statements* are theory-laden, implying that data interpretation depends on theoretical frameworks [4]. Popper rejects the idea that induction can distinguish science from non-science, as it cannot justify theories. Instead, he proposes *falsifiability* as the *demarcation* criterion: a theory is scientific only if it can, in principle, be falsified by empirical evidence. When tests fail to disprove the theory, it is *corroborated*, though never conclusively validated [5].

Corroboration is Popper's technical term for when a hypothesis passes empirical tests designed to falsify it. It indicates temporary empirical support of the hypothesis but not its confirmation or truth. Significantly, Popper argues that corroboration does not, and cannot, make a theory *probable.* Thus, even regularly corroborated theories remain hypothetical and subject to future falsification.

As emphasised in [5], no matter how extensive, corroboration carries no evidential weight in Popper's framework. A corroborated hypothesis is one that (1) has not yet been refuted and (2) has survived severe tests designed to falsify it [6]. Crucially, even theories that survive all challenges never attain probability; they remain hypothetical. Conversely, a single negative test result decisively *falsifies* the theory.

Engineers and scientists often treat repeated corroboration as evidence that a theory or design is provisionally true. However, Popper emphasises that corroboration never proves the truth; it only indicates that the theory has survived severe tests. This reveals his key insight about the fundamental *asymmetry* between verification (logically impossible for universal claims) and falsification (decisive in principle).

While corroboration cannot verify a theory's truth, Popper acknowledged that falsification is not automatically decisive *in practice.* While corroboration cannot verify a theory's truth, Popper emphasises that falsification is logically decisive for universal claims. A single valid counterexample can refute them conclusively in principle. However, he acknowledges that in practice, falsification is not automatically decisive, and a single contrary observation may not rationally demand revision if:

- The experimental result is questionable, or
- The theory can be revised and modified through *non-ad-hoc* modification.

However, this is distinct from Poppe's logical asymmetry: falsification can be conclusive for universal theories if the counterevidence is valid. In practice, engineers may retain falsified theories temporarily when alternatives are lacking—but this reflects *pragmatic constraints*, not *theoretical validity*.

Also, according to Popper's perspective, in scientific theories, a theory's value is determined by its level of *informative content*, *testability*, and the *relationship* between these factors and its *probability*.

The informative content of a theory refers to the extent of new knowledge and explanatory power it offers. However, as a theory's probability increases, its informative content tends to decrease. Note that in the context of the philosophy of science, a theory's probability refers to the likelihood or chance that the theory accurately describes and explains a particular phenomenon or set of observations. In simpler terms, highly probable theories might lack detailed specificity and be more general. Such highly probable statements might even be trivial or widely known, resulting in limited novelty or fresh insights.

Popper also asserts that the informative content of a theory is directly related to its testability. A theory that can be rigorously tested and potentially falsified provides more opportunities for generating new information and insights. Testable theories can be put to empirical tests, and the outcomes of these tests contribute to the informative content.

Popper's view suggests that scientific theories should be valued not solely based on how likely they seem to be true but on their ability to provide new insights and withstand empirical testing. This perspective encourages the pursuit of theories that are specific, informative, and open to empirical verification or falsification, which aligns with Popper's emphasis on falsifiability as a hallmark of scientific theories.

7.3.2 The Concept of Theory-Laden Observations

In philosophy, the term *theory-laden* refers to the idea that our observations and perceptions are influenced or shaped by our theoretical framework, prior knowledge, or conceptual assumptions. This means our understanding of the world is not purely objective but filtered through our existing theories and frameworks. When scientists or engineers conduct experiments or make observations, they do so within the context of existing scientific theories. These theories guide what they choose to observe, how they interpret the data, and what conclusions they draw. The idea of theory-laden observations in philosophy can be attributed to several influential philosophers, including Immanuel Kant, Thomas Kuhn, Willard Van Orman Quine, Paul Feyerabend, and Karl Popper, who significantly contributed to its development.

The perspective on observation being theory-laden is a concept that challenges the common notion that observations are neutral and independent of one's preconceived ideas or theories. This implies that our perceptions are filtered and biased through our existing theories, beliefs, and expectations when we observe a phenomenon. Therefore, we do not observe raw, unfiltered reality; our observations are influenced by the theories we are already familiar with.

The theory-laden concept has been interpreted and discussed in various ways within the philosophical and scientific communities. The two key interpretations relevant to our control engineering framework are as follows:

- A *strong* interpretation suggests that our theoretical frameworks determine observations and cannot be separated from them. According to this perspective, our theories influence how we interpret and understand our observations and what we perceive in the first place. This interpretation is often associated with Thomas Kuhn's work on scientific revolutions and paradigms.[1]
- A *weak* interpretation argues that theories do not entirely determine observations but are still influenced by them to some extent. In this view, observations can provide some degree of neutral information that can challenge or modify existing theories. This interpretation is more aligned with the ideas of Karl Popper and his emphasis on falsifiability and critical testing of theories.

It will be argued later in this chapter that the weak interpretation stance is more applicable in engineering design applications where the goal is often to develop practical solutions based on empirical evidence. The raw and pure data obtained via a sensory system without any prior filtering are neutral information that can be used in different design frameworks. In control systems design, the data would be the actual manipulated inputs and measured outputs of the plant. The set of inputs and outputs is the closest we can get to understand and realise the behaviour of a system. Note that the source of the manipulated input is irrelevant when the system's behaviour is of concern, and what actually matters in engineering is how the system reacts to these inputs. However, these data are further utilised for system analysis and modelling from a particular theoretical framework, and the subsequent observation will become theory-laden.

Our theories guide us in selecting *what* to observe and *how* to interpret it. When studying a phenomenon, we tend to notice and highlight things that align with our existing theories while ignoring or downplaying observations that contradict them. This can lead to unconsciously favouring observations that confirm our theories.

Observation involves interpretation based on our existing knowledge and is not merely a passive recording of information. This interpretation is strongly influenced by the theoretical frameworks accepted by the observer. However, testing theories against reality would be straightforward if observations were entirely detached from theories. Since observations are influenced by theories, the process of falsification becomes more intricate. A theory's rejection based on observations can be influenced

[1] This is Kuhn's paradigm-dependence of observation, where scientists working in different paradigms may *see* different things in the same data and implies incommensurability, where observations are so theory-laden that competing paradigms cannot be neutrally compared.

by the theory and the observer's theoretical biases. Let us elaborate on the notions of observational and theoretical.

Observational refers to system investigation aspects involving direct empirical observation, measurement, and data collection from the system. Observational data are obtained through sensory experience, experimentation, or measurement and are considered the foundation of the *behavioural* study of systems. They provide the basis for testing hypotheses, constructing theories, and making predictions.

Theoretical, conversely, pertains to concepts, models, explanations, and frameworks constructed to interpret, explain, and predict observed phenomena. Theoretical components often involve abstract reasoning, mathematical formalism, and conceptual frameworks that go beyond direct empirical observation. Theories are constructed to organise and make sense of observational data and explain why certain phenomena occur.

To start a practical discussion in line with control systems theory, consider the observation of a pendulum's motion. If the pendulum's period is measured for different lengths, the data collected from these measurements are empirical observations. These observations serve as the foundation for understanding the relationship between the pendulum's length and its period. The empirical data collected through direct measurements are observational in nature. Now one may notice a pattern in the data; that is, longer pendulum lengths seem to result in longer periods. To explain this pattern and make predictions beyond the collected data points, a mathematical model can be developed for the pendulum. The model would consist of differential and algebraic equations describing why the pendulum's length affects its period. This theoretical model goes beyond direct observation and explains the observed phenomenon.

In this simple example, the observational aspect involves measuring the pendulum's periods for different lengths, while the theoretical aspect involves constructing a mathematical model to show the relationship between length and period. Based on these observations and theories, the pendulum behaviour analysis is generalised beyond the specific pendulum. This provides a platform to predict how the pendulum will behave under different conditions.

Note that the observations themselves are generally considered to be relatively theory-independent at the moment they are made. Theoretical considerations do not inherently influence the direct measurement of the period of a pendulum for various lengths. However, the way we interpret and analyse those observations can become theory-laden. In other words, the raw observations might not be theory-laden, but making sense of those observations, recognising patterns, explaining mechanisms, and connecting them to established scientific theories involves presupposing our existing theoretical knowledge. So, while measuring the pendulum's period and length is not theory-laden, interpreting what those measurements mean in terms of physics and the relationships between variables involves theoretical considerations.

To be more specific in the control theory domain, consider the terms and predicates like *faster*, *stable*, *overshoot*, and *oscillation* in control system design. These terms derive their meaning directly from hands-on experiments and practical experiences. Consider, for instance, James Watt's observation of the responses of the fly-ball governor. Even before developing mathematical formulations and system theories by Maxwell [7], he possessed an intuitive grasp of these terms, as they corresponded to the behaviours exhibited by his system. The categorisation of responses as faster or slower, responses leading to an increasing output amplitude resulting in the phenomenon we now term *instability* (referred to as *haunting* in the context of the governor), and those governor responses that avoid excessive oscillations or divergence over time, identifiable through empirical observations, all contribute to what we now label as *stability*. Similarly, when the output amplitude surpasses a predetermined value, it is referred to as an *overshoot* in modern terminology, and instances of *oscillations* in the governor response further illustrate these concepts.

Our understanding of the governor's behaviour mirrors that of James Watt. Any observer, devoid of technical-mathematical knowledge, could have intuited these terms or their equivalents. However, investigating and analysing these responses presents a separate perspective and can indeed be laden with theoretical considerations, as was evident after Maxwell's mathematical analysis of the governor. In conclusion, the measurements and observations revealed the system's behavioural performance and can confidently be classified as observational.

On the other hand, control theory employs theoretical models and mathematical formulations to gain a deeper understanding of system behaviour, analyse performance, and design controllers. Theoretical concepts and models enable control engineers to predict outputs, conduct simulations, and devise strategies for achieving desired system behaviour based on these mathematical models and theories. In contrast to the behavioural properties discussed earlier, these are *representational* characteristics of the system. Theoretical terms such as *state controllability*, *optimality*, *internal stability*, *Lyapunov stability*, and *state-space realisation* in control systems design derive their significance from specific mathematical theories of systems. Researchers operationalise theoretical terms via observational counterparts where feasible (e.g., stability to bounded physical outputs). However, many concepts (e.g., controllability, Lyapunov stability) derive meaning primarily from mathematical frameworks, with only indirect empirical validation [6].

State-controllability refers to the property of the state-space mathematical model of a system and determines whether it is possible to steer the system's state variables from an initial state to a determined final state using available control inputs. This concept is rooted in the mathematical theory of control systems and is not necessarily a behavioural or intrinsic characteristic of the actual system.

Similarly, optimality is a mathematical property that can be achieved by optimising certain defined cost functions. Different results will be achieved depending on the cost function and the optimisation technique employed.

Internal stability relates to a property of the systems' mathematical description of its internal dynamics, independent of any external inputs or outputs. Analysing internal stability requires mathematical techniques from control theory, such as analysing right half-plane pole-zero cancellations. The important concepts of poles and zeros in model-based control are theory-laden and are not observational. Measuring the poles and zeros of an actual physical plant is not feasible; instead, these parameters are defined within a mathematical model that relies on a specific theoretical background in system theory.

State-space realisation involves representing a system's mathematical description in terms of its state variables, inputs, and outputs. It is considered a fundamental concept in state-space control theory and has no direct counterpart or consequence in the behaviour of a system.

These terms are theoretical because they draw specific meanings from formal mathematical frameworks. They are used to analyse and design control systems based on mathematical models and principles rather than direct empirical observations at a representational level. The essence of model-based control strategies is these theory-laden observations and different theoretical frameworks. However, truly data-driven control techniques should be based on observations and measurements. These cases are depicted in Fig. 7.1. In Fig. 7.1a, raw data is measured and collected for model-free control, which is then utilised without further manipulation in a data-driven control systems design framework. In contrast, as depicted in Fig. 7.1b, information, assumption, and raw data undergo manipulation within a mathematical system theory framework. This manipulation establishes theory-laden observations, such as poles and zeros, controllability and observability, eigenvalues and eigenvectors, uncertainty upper bounds, and so forth.

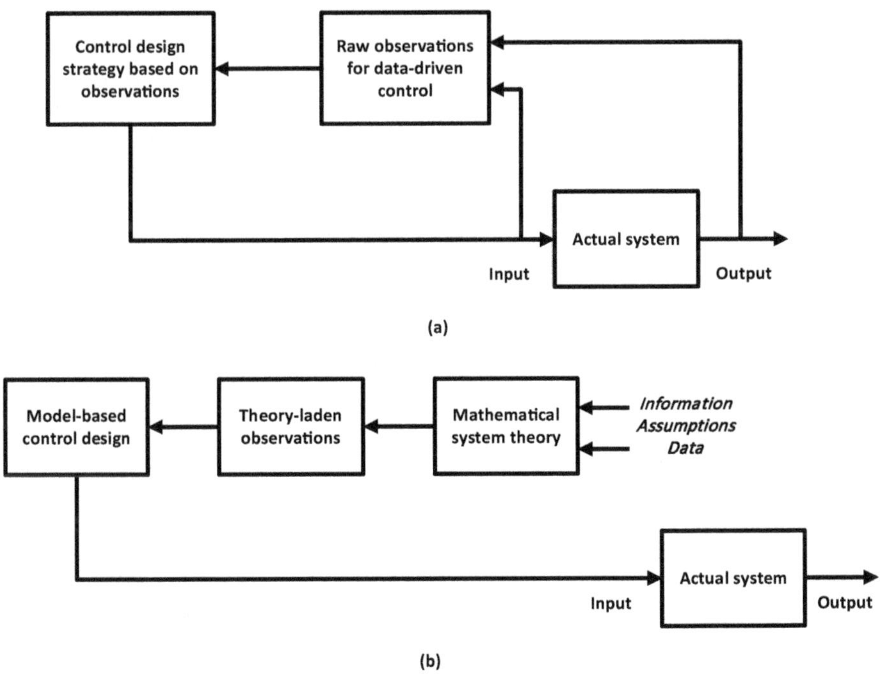

Fig. 7.1 **a** A control design strategy based on the raw observational data. **b** A control design strategy based on the theory-laden observations

7.3.3 The Problem of Infinite Regressors

The problem of infinite regress arises when employing induction-based reasoning. In this regime of reasoning, each *conclusion* depends on prior *assumptions*, which rely on further assumptions, creating an endless chain of justification. This creates infinite sequences of units governed by a *recursive principle*. Every unit is governed by the same dependency principle, leading to a regression without a definitive foundation.

Flammarion's Sphere or *Flemming's Turtle* is a famous illustration symbolising infinite regression and the limits of human understanding. It represents a terrestrial sphere balanced on a turtle, which endlessly stands atop another turtle. This concept is known as *Turtles All the Way Down*. This metaphor mirrors the problem of infinite regress: each turtle represents a dependent unit in the sequence, relying on its predecessor for justification without ever reaching a foundational layer. The endless recursion underscores the challenge of explaining complex phenomena, as every assumption demands further explanation without any ending.

A key example of infinite regression arises in Bayesian probability and its estimation methods [8]. Bayes' theorem defines conditional probability as:

$$P(A|B) = \frac{P(B|A)P(A)}{P(B)} \tag{7.1}$$

where, the posterior probability $P(A|B)$, the probability of parameters A given evidence B, depends on the prior probability $P(A)$. However, determining $P(A)$ itself often requires another prior, creating a recursive dependency similar to the turtles all the way down problem. Since each prior depends on a previous prior, the chain of dependencies extends infinitely unless terminated by assumption, typically through uninformative priors.

Another example of infinite regression appears in *recursive identification techniques*, such as those used in adaptive control strategies. The widely known technique, the *Recursive Least Squares* (RLS) algorithm, iteratively estimates dynamical system parameters from input–output data but requires initial parameter guesses to begin the process. The selection of these initial estimates can be crucial, as they affect the convergence and accuracy of the algorithm. A solution to this problem is to initialise RLS using estimates from another identification technique. However, this secondary method may itself depend on prior estimates, creating a chain of recursive dependencies. This leads to an infinite regression problem: the accuracy of each stage hinges on the preceding one. If any of the initial estimates are inaccurate, the inaccuracies propagate through the subsequent recursive stages, potentially leading to poor overall parameter estimation.

This scenario can be replicated for all other tuning parameters in analogous recursive identification and control methodologies. In other words, the requirement for precise initial parameter estimates at each stage can potentially result in an infinite regression issue if each subsequent estimation approach depends on preceding estimates obtained through varying techniques.

Moreover, the foundational assumptions in mathematical models of physical systems regarding their structure or properties are not derived directly from the systems themselves. Instead, they often stem from the mathematical prerequisites of a particular design theory and are validated against approximate mathematical representations of reality. This approach inherently conflicts with Newton's *hypothesis non fingo principle*,[2] which rejects unfounded assumptions in scientific explanation.

Falsification can serve as a response to the problem of infinite regress. Requiring falsifiable predictions prevents an endless chain of ad hoc explanations. This approach helps to avoid the problem of *circular justifications*[3] by demanding the possibility of being proven false through observation. This is achieved in a data-driven black-box methodological approach, where only observed data are assumed available, without any prior assumptions about the properties or structures of the *actual system* or measurement noises [8]. By demanding that predictions can be proven false through observation, designers can continuously refine their understanding of the system and improve their results. This iterative process helps to avoid circular justifications and promotes a more robust and reliable practical engineering understanding of the system being studied.

The requirement of empirical testing inherent in falsification provides a concrete methodology for breaking the cycle of infinite regress in validating scientific theories. Instead of relying on endless justifications and recursions, designs can be subjected to rigorous empirical testing, providing a practical way to avoid an unfounded chain of reasoning.

7.4 Falsification Alternative to Induction-Based Feedback Control Philosophy

7.4.1 The Model-Based Control Philosophy

The model-based control philosophy relies on *inductive reasoning* rather than *strict falsification*. Model-based control starts with constructing mathematical models that describe the dynamics of a system. These models are derived from observations,

---

[2] This principle reflects Newton's approach to scientific investigation that signifies the preference for grounding theories and conclusions in empirical observations and rigorous mathematical deductions rather than speculative hypotheses or conjectures. It is a Latin phrase that translates to *I do not create a hypothesis* or *I put forward no hypothesis*. Newton employed this phrase to indicate that his work was focused on presenting the laws of motion and universal gravitation as established facts, rather than proposing hypotheses or speculative theories. In a broader context, it reflects an approach where the researcher aims to provide well-supported, empirical evidence for their findings, without relying heavily on theoretical assumptions.

[3] Circular justification, also known as *circular reasoning* or *circular argument*, takes place when the assertion being attempted to be proven is incorporated as either a constituent of the reasoning process or as one of the premises in the argument.

experiments, and empirical data. This process involves *inducing* patterns and relationships from available information to create a representation of how the system behaves.

Once a model is constructed, it is used to predict the system's behaviour under different conditions. Control strategies are designed based on these predictions. When the control strategy successfully guides the system's behaviour in accordance with the model's predictions, it confirms the validity of the model in those particular scenarios. This confirmation is analogous to *inductive reasoning*, where observed patterns reinforce the validity of a general hypothesis.

While model-based control may involve some degree of falsification when predictions do not match observations, the emphasis is often on refining the model to better capture the system's behaviour rather than outright rejecting the model. Unlike strict falsificationism, where a single disconfirmation can lead to the rejection of a theory, model-based control might involve incremental adjustments and modifications to the model to align it more closely with observed behaviour.

In this alignment process the goal is to build up a better understanding of the system over time through repeated observations and adjustments. Thus, model-based control fundamentally relies on inductive reasoning. It constructs system models from observed data and iteratively refines them, symbolising the inductive process of generalising from specific observations to broader predictive frameworks.

Consider, for instance, the challenges in robust control synthesis methods discussed in the context of model-based design [9]. To initiate a robust control design strategy, it is necessary to determine an upper bound for plant model uncertainty. This is a key assumption in all closed-loop robust stability and performance proofs. Hence, different identification techniques are utilised to provide both a nominal model and an upper bound for model error. This typically involves a process of refinement of the model and bounds that are consistent with inductive reasoning, and the designer attempts to derive a more appropriate nominal model and accurate bounds over time through repeated observations and adjustments. An early example of this strategy is presented in [10], where it is proposed to modify the identification process such that along with estimating model parameters, a measure of model uncertainty is also obtained. Then, with such an uncertainty estimate, a robust controller is designed. However, if the resulting controller is too conservative because the model estimate is too uncertain, then it is necessary to repeat the estimation procedure to obtain a better model. On the other hand, in any identification scheme, a priori assumptions about the unknown system, the noise, and the model set from which the identified model is chosen are assumed.

However, the presence of *unverifiable* assumptions about model accuracy can impact the effectiveness of establishing upper bounds for model error. Experimental data inherently captures only a limited subset of plant behaviours, establishing merely a lower bound on the uncertainty that is insufficient to determine the conservative upper bound required for robust control guarantees. Since robust stability and performance proofs require a valid upper bound on plant uncertainty, which experimental data cannot provide, the theoretical guarantees of robust control fundamentally cannot be realised in practical implementations. Consequently, no rigorous

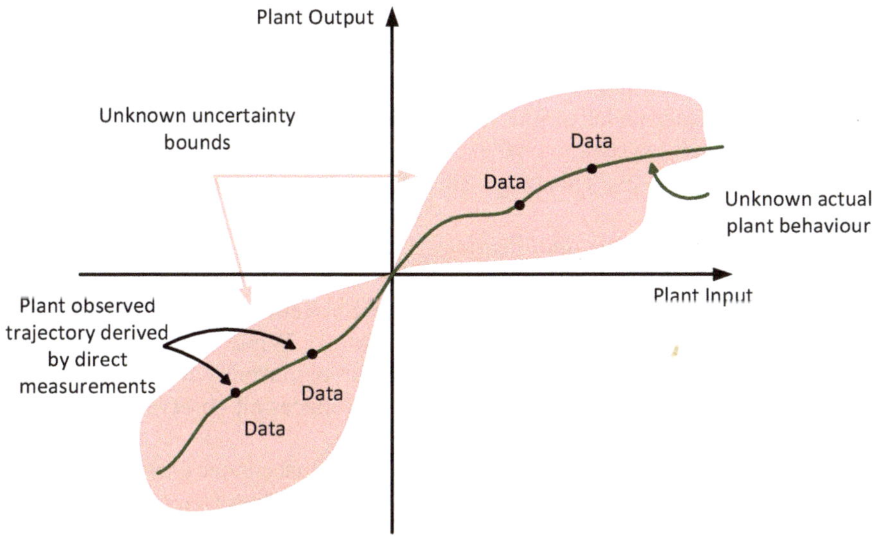

Fig. 7.2 The known plant trajectory and unknown required uncertainty bound

performance assurances can be made for real-world systems. This necessitates a substitute for guaranteed performance. This is depicted in Fig. 7.2 [9]. The plant input–output observed trajectory is the only available raw data that is shown as a graph in the input–output space. In the case of uncertain systems, a rather larger subset of input–output space must be considered. This larger set is characterised by the unknown uncertainty upper bounds. Hence, achieving a guaranteed performance based on an exactly derived upper bound is impossible with an induction-based control philosophy, as it relies on the aforementioned unverifiable assumptions.

Finally, achieving performance guarantees demands that controllers, past data, and design specifications remain mutually consistent, which is a requirement achievable via falsification-based control approaches. It will be shown in Sects. 7.4.2 and 7.5 that the falsification-based control philosophy provides this consistency without any additional assumptions and solely utilising observational raw data.

7.4.2 Falsification in Control Systems

The principles of falsification philosophy have impacted the field of control systems. Falsificationist principles could be applied to examine models and subject them to rigorous testing and validation against real-world data. This helps identify discrepancies between the model predictions and actual system behaviour, leading to more accurate and reliable control strategies. Also, falsificationist philosophy encourages control engineers to design controllers that can handle a range of scenarios, including worst-case scenarios. Controllers are designed and tuned to be robust in the face of uncertainties and disturbances.

Control systems are typically tested using simulations and experiments to evaluate their performance under various conditions. A falsificationist approach involves subjecting the control system to challenging scenarios that test its limits and potentially expose weaknesses. This helps identify weaknesses and refine the control strategy.

Falsification in systems identification. A concept that is close to falsification is invalidation, which some researchers follow. An early example appears in [11], which notes that its applicability to the physical system must be evaluated after determining a model (even through ad hoc methods). While engineers can never guarantee that the model describes all possible input–output behaviours, they can test a necessary condition to check that the model reproduces all observed behaviours. This constitutes the model *invalidation problem.*

Invalidating generally refers to making a model inaccurate, unreliable, or not representative of the actual system dynamics. It often involves factors that make the model unsuitable for describing or predicting the behaviour of the real-world system. It is commonly used to describe the process or factors that lead to the failure of a model to represent the system under consideration accurately. However, falsification is a broader concept, subjecting a model to tests and experiments to find conditions under which it fails to predict or describe the system accurately. In system identification, falsification is often employed to test models rigorously. The process involves subjecting the model to various conditions and data sets to determine its limitations and ensure that it represents the system.

The concept of falsification was first formally introduced and operationalised in system identification by Jan Camiel Willems [12]. For the first time, the ideas and language used in data modelling were evocative of Popper's philosophy of science. The guiding principle here is the dictum that *the more a model forbids, the better it is.* This principle suggests that greater restrictions or limitations in a model contribute to enhanced explanatory and predictive power. Specifically, a model should avoid being overly broad or general, as this may compromise the specificity and precision required for accurate predictions.

According to the falsification philosophy, as suggested by Willems, a model remains *unfalsified* if it incorporates all relevant information available up to a specific moment. This underscores the significance of integrating comprehensive and up-to-date data into the modelling process. The identified or developed mathematical models to represent real-world systems must be consistent with the available data and observations.

Conversely, a model is said to be *falsified* when it fails to accurately capture the behaviour of the real system as reflected in the observed data, and an unfalsified model is one that has not been contradicted or invalidated by the available observed data up to a certain point in time. In other words, it is a model that successfully incorporates all the relevant information from the system at that given moment. The concept of falsifying and unfalsified models helps researchers and engineers understand the degree to which a proposed model aligns with real-world observations.

Additionally, the *Most Powerful Unfalsified Model* (MPUM) concept refers to the model that best fits the available data. In the context of model comparison, multiple models can be consistent with the available data. However, some of these models might be simpler or more complex than others. The MPUM is the model among the candidate models that provides the best fit to the data while still being unfalsified.

In other words, among the unfalsified models that are considered, the MPUM is the one that maximises the likelihood of observing the collected data. It strikes a balance between model complexity and data fit, and it is a practical approach to selecting a model that is both informative and parsimonious. Based on the falsification theory, an exact definition and an algorithm for the computation of MPUM based on system observations is given in [13].

The MPUM is formulated within the context of behavioural control modelling. Let us assume a phenomenon we wish to model, expressed mathematically in the behavioural control modelling framework. We designate the phenomenon's outcomes as elements in a set $\mathbf{U}$ which is called the *universum*. Elements of $\mathbf{U}$ Will be called the outcomes of the phenomenon. A deterministic mathematical model viewed purely from a behavioural or black box perspective asserts that certain outcomes are possible and defines a subset $\mathbf{B}$ of $\mathbf{U}$ as the behaviour. Hence, a mathematical model is represented as a pair $(\mathbf{U}, \mathbf{B})$ with the universum $\mathbf{U}$ and behaviour $\mathbf{B}$. Let $\mathbf{M}$ be a set representing a family of mathematical models. A model $(\mathbf{U}, \mathbf{B})$ is defined unfalsified by data $\mathbf{D} \subseteq \mathbf{U}$ if $\mathbf{D} \subseteq \mathbf{B}$. Also, a model $(\mathbf{U}, \mathbf{B}_1)$ is defined as more powerful than $(\mathbf{U}, \mathbf{B}_2)$ if $\mathbf{B}_1 \subseteq \mathbf{B}_2$ aligning with the dictum that the more a model forbids, the better it is. Now, M is the MPUM in $\mathbf{M}$ based on the data $\mathbf{D}$ if the following conditions hold:

i. $\mathrm{M} \in \mathbf{M}$
ii. M is unfalsified by $\mathbf{D}$
iii. If $\left\{\mathrm{M}' \in \mathbf{M}, \mathrm{M}' \text{ unfalsified by} \mathbf{D}\right\}$ gives $\{\mathrm{M} \text{ is more powerful than} \mathrm{M}'\}$.

Falsification in control systems design. Predefined mathematical assumptions and a priori system knowledge constitute the Achilles' heel of induction-based control system design methodologies. Refer to Sect. 3.3.5 for more details. Principles of the falsification philosophy can be employed to develop an assumption-free methodology using system observations, which are pure, raw, noise-corrupted, past input–output measured data. Such an approach would constitute a *truly* data-driven methodology.

The falsification theory offers the necessary foundation for an alternative to the commonly and widely used inductive-based methodologies. Based on falsification, the deductive-based approach to control system design emerges. In control systems design and within the falsification process, it is essential to subject all *hypotheses* to rigorous testing. Hypotheses indicative of closed-loop stability and performance are formulated and tested through observations and experiments to determine their *validity*. This methodology of falsification is often referred to as the *hypothetico-deductive* method. It is an alternative to the approach that traditionally derives knowledge from analogy and classical reasoning (Aristotelian-Platonic).

Furthermore, the falsification approach has been incorporated into methodologies of data-driven control theory. Rigorous testing involves either prediction or experimentation. Prediction entails extrapolation from the system's current state, whereas experiments function as verifications and investigations designed to validate the relationships within the system. Predictions that diverge from the test results are subsequently falsified or refuted.

The falsification theory encompasses three key features crucial to control systems design:

- *Falsifiability*: Theories can be falsified if new observational data (from experiments) contradicts the theory.
- *Parsimony*: The theory should remain simple and not require additional assumptions beyond the observed data's required assumptions.
- *Validation*: The theory must undergo thorough validation through rigorous experiments to identify potential counterexamples.

Practising the aforementioned principles in control systems design involves defining an appropriate performance criterion. Subsequently, a candidate controller must be chosen from the many potential controllers based on this criterion. The final controller selection is made solely using the measured actual plant data to be placed within the feedback loop.

However, the control system can only access a finite record of plant input–output data, representing past system interactions and responses. This inherent data limitation constrains the system's ability to predict future plant behaviour, as the available dataset cannot encompass all possible operational scenarios. Hence, to predict the plant's behaviour in response to future signals, the control system often needs to make certain assumptions, as a limited number of these future signals might resemble signals that the control system has previously encountered. In most cases, these signals will be entirely new.

Hence, we are faced with a scenario in which the act of prediction becomes crucial for generating the forthcoming control signal. A critical constraint in this framework is the prohibition against using system models or any additional representational assumptions when making these predictions.

Falsificationist principles can address this problem. In this framework, a controller's definitive validation is its unfalsifiability, which is its consistency with all available empirical data up to the present. When a controller withstands rigorous falsification attempts and aligns with observed measurements, it achieves *corroboration*. By falsificationist logic, data-driven control theory is inherently constrained to performance verification using only the available pure data. Since no model assumptions are permitted, the theory can only test whether the controller's behaviour remains *unfalsified* by the empirical record.

However, the model-based approach for closed-loop stability proofs rests upon a foundation of assumptions. In contrast, the falsification approach operates without a preceding assumption and operates solely under the conditions of modest *data informativity* and *cost detectability*. In this manner, unsuitable controllers are identified and rejected through falsification without any predetermined assumptions.

Data informativity[4] is a data-driven analysis and control framework where an unknown system generates observed data. Without relying on prior parametric models, the framework evaluates whether these data are sufficiently informative to achieve specific control or analysis objectives, such as system identification, stabilisation, or robust controller design. The objective is to evaluate the system-theoretic properties of the actual system and design control laws using only measured data. This problem can be formulated as follows. Given a target system-theoretic property and the class of all systems exhibiting this property, we aim to determine whether the actual system satisfies the property. The sole available information consists of empirical observations. A conclusive verification can be made if every system consistent with the observed data necessarily possesses the property. This motivates the definition of data informativity: the measured data are informative for the property if the set of all data-consistent systems is contained within the set of systems satisfying the desired property.

Cost detectability, as defined in the context of unfalsified adaptive control, relates to the performance criterion (cost function) and controller characteristics. This property is independent of any plant dynamics or signal assumptions. Its role is to guarantee that, for every candidate controller, boundedness of the cost function provides both necessary and sufficient conditions for stabilizability, or equivalently, it serves as a necessary and sufficient condition for *corroboration permit*.

7.5 Falsification in Action

Building on Willems' falsification concepts in system identification, Safonov pioneered in a 1994 paper unfalsified control by recognising that input–output data could validate control laws with performance bounds as directly as it validates plant models with uncertainty bounds. This insight enabled a less conservative, more immediate approach to robust control design by bypassing traditional plant modelling steps [9].

Unfalsified Adaptive Switching Control (UASC) algorithms seek to identify a stabilising controller from input–output data with minimal assumptions [14]. Following Safonov's 1995 introduction of falsifiability to control system design [15], UASC employs a data-driven supervisor rather than system identification. While no explicit system model is used in the UASC process, the controller bank (a pre-designed set of candidate controllers) may incorporate model-based designs.

This model-free approach can handle indeterminate systems. As shown in Fig. 7.3, the supervisor selects controllers from the available set based on real-time system data, inserting them into the control loop. The method assumes the controller bank contains at least one *feasible* candidate.

[4] A conceptually related requirement in model-based control design is the notion of *persistency of excitation*, which is a condition ensuring unique identifiability of dynamical systems within a specified model class.

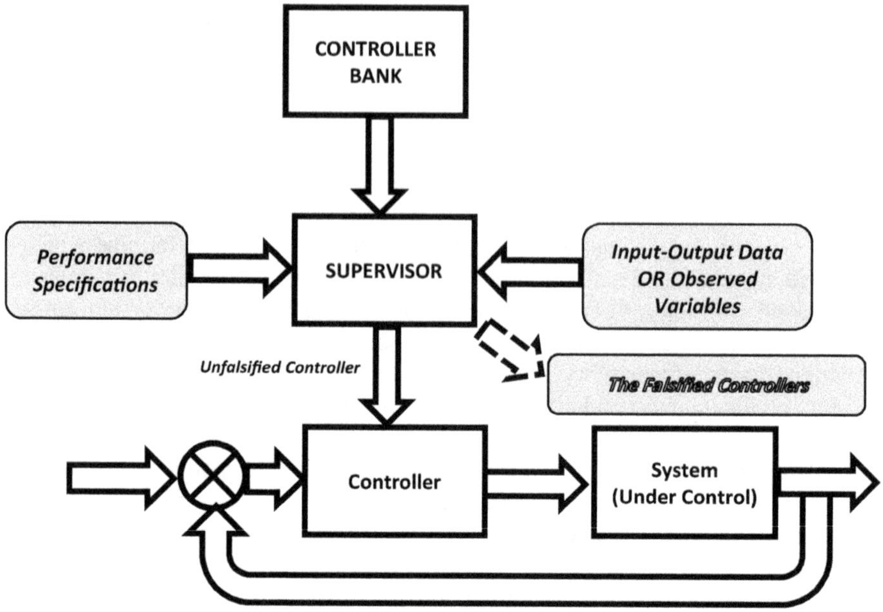

Fig. 7.3 Concept of the controllers' falsification

As shown in Fig. 7.3, it is assumed that measured input–output data and a predefined set of controllers with known performance specifications are available. A controller is considered unfalsified if it:

1. Maintains system stability according to the observed data, and
2. Satisfies the prescribed performance criteria.

Conversely, a controller is falsified if the empirical data demonstrates either:

1. Failure to stabilise the system, or
2. Inadequate performance.

Note that this evaluation process occurs without implementing controllers in the closed loop, and the controllers are falsified solely based on input–output data analysis. Safonov and co-workers propose the principles of unfalsified adaptive control in a series of technical papers.

The controller set consists of both an active controller currently operating in the closed-loop system and multiple inactive candidate controllers. The supervisor evaluates controller performance in two distinct ways: for the active controller, it uses direct online measurements, while for inactive controllers, it employs the virtual reference signal approach and the observed data. This innovative method generates virtual reference signals derived exclusively from the system's input–output data and each controller's structure.

Each inactive controller has an associated virtual reference signal—a theoretical input that would have produced the same observed input–output data up to that time had that particular controller been active in the loop. The performance assessment uses a specialised cost function that compares controllers' performances solely based on the observed data. This cost function calculation incorporates weighted measures of both tracking error and control effort, normalised by reference signal magnitude to ensure fair comparison across different operating conditions.

The virtual reference signals can be computed directly from the input–output data and controller structure, requiring no system model. This computation simplifies to an algebraic relationship combining the controller's inverse with the measured signals for minimum-phase proper controllers. A key requirement is feasibility—the controller set must always contain at least one stabilising controller to ensure the adaptive control problem remains solvable.

This falsification-based approach represents a novel control design strategy. Eliminating dependence on mathematical models and instead using direct data-driven evaluation enables control implementation for complex systems where traditional modelling is impractical or impossible. The method's deductive nature—systematically eliminating underperforming controllers rather than inductively building models—provides a robust framework for real-world applications with uncertain dynamics.

7.6 Conclusions

The falsification concept introduced in the philosophy of science is briefly reviewed. The falsificationist principles have influenced engineering and applied sciences, particularly control systems engineering. These principles have influenced system identification, modelling, and controller design methodologies. Applying falsification theory has led to a practical data-driven control system as one of the emerging data-driven control methodologies. It is shown that the falsification mindset frees the control system designers from the induction-based control systems design methodologies.

References

1. K. Popper, *The Logic of Scientific Discovery* (Routledge, 2005)
2. A. Khaki-Sedigh, *An Introduction to Data-Driven Control Systems* (Wiley-IEEE Press, 2023)
3. M. Staples, Critical rationalism and engineering: methodology. Synthese **192**, 337–362 (2015)
4. T. Lewens, *The Meaning of Science: an Introduction to the Philosophy of Science* (Hachette UK, 2016)
5. B.J. Caldwell, Clarifying popper. J. Econ. Lit. **29**(1), 1–33 (1991)
6. S. Psillos, *Philosophy of Science AZ* (Edinburgh University Press, 2007)
7. J.C. Maxwell, On governors. Proc. R. Soc. Lond. **16**, 270–283 (1868)
8. M.G. Safonov, Robust control: Fooled by assumptions. Int. J. Robust Nonlinear Control **28**(12), 3667–3677 (2018)

9. M.G. Safonov, T.-C. Tsao, The unfalsified control concept and learning, in *Proceedings of 1994 33rd IEEE Conference on Decision and Control*, vol. 3 (IEEE, 1994), pp. 2819–2824
10. R.L. Kosut, Adaptive uncertainty modeling: on-line robust control design, in *1987 American Control Conference* (IEEE, 1987), pp. 245–250
11. R.S. Smith, J.C. Doyle, Model invalidation: a connection between robust control and identification, in *1989 American Control Conference* (IEEE, 1989), pp. 1435–1440
12. J.C. Willems, Paradigms and puzzles in the theory of dynamical systems. IEEE Trans. Autom. Control **36**(3), 259–294 (1991)
13. I. Markovsky, J.C. Willems, S. Van Huffel, B. De Moor, *Exact and Approximate Modeling of Linear Systems: a Behavioral Approach* (SIAM, 2006)
14. M.G. Safonov, Origins of robust control: early history and future speculations. Annu. Rev. Control. **36**(2), 173–181 (2012)
15. M.G. Safonov, T.-C. Tsao, The unfalsified control concept: a direct path from experiment to controller, in *Feedback Control, Nonlinear Systems, and Complexity* (Springer, 1995), pp.196–214

Chapter 8
Paradigm Shifts in Control Theory: From Kuhn's Perspective

To reject one paradigm without simultaneously substituting another is to reject science itself.
—Thomas S. Kuhn

8.1 Introduction

Researchers from various domains have utilised the philosophy of science to investigate their fields of expertise for a deeper understanding and to enhance both theory and application. Refer to [1] and the references therein for specific examples. In utilising the tools and concepts of the philosophy of science in the engineering domain, it is important to acknowledge that the nature of engineering differs from science, and theories in engineering are different to theories in science. Hence, the growth of knowledge in engineering is different from that in science.[1] These discrepancies should be accounted for when applying the concepts and methodologies from the philosophy of science. However, the established frameworks in the philosophy of science can be adapted to study the methodological issues in the epistemology of engineering with some considerations in this mapping. Another key issue in studying engineering systems is the analysis of the sources of error in engineering or engineering failures. Subsequently, the responses to such failures can lead to the growth of knowledge in engineering. Evolutions follow this in the particular school of thought in engineering or can lead to fundamental changes or adoption of radically different approaches, called *revolutions* in engineering knowledge.

The philosophy of science encompasses various perspectives on defining and understanding scientific practices. One of the significant contributions to this field is Thomas Kuhn's theory of *scientific revolutions*, which introduced the concept of *paradigm shifts*. This theory offers profound insights into the dynamics of scientific advancement and the transformation of knowledge over time.

[1] Refer to Sect. 1.3 for engineering ontology with emphasis on interconnected and divergent concepts and methods for engineering and science. Also, refer to Sect. 7.2 for remarks on engineering knowledge and responding to failures in engineering.

A. Khaki-Sedigh, *Control Systems: A Historical and Philosophical Perspective*,
https://doi.org/10.1007/978-981-96-7426-8_8

Although Kuhn's approach and scientific revolutions have been primarily developed and used in studying the natural sciences, they can be potentially valuable for engineering applications. In engineering, paradigms are the dominant ways of thinking, problem-solving, and designing that guide the profession. Historical developments, technological advancements, and the broader cultural and societal context often influence these paradigms. They can shape how engineers approach problems, the tools and techniques they use, and the criteria they employ to evaluate the success of their solutions.

Two general examples of engineering paradigm shifts include the transition from craft-based to scientific and systematic engineering practices during the Industrial Revolution. This change led to a more standardised and quantifiable approach to engineering. Furthermore, the *digital transformation* has resulted in paradigm shifts in engineering, enabling the design and analysis of complex systems while replacing traditional manual methods.

In this chapter, we shall apply Kuhn's paradigm shift model to analyse control systems design evolution, and the rationale for this is as follows:

- The historical progression of control design (from antiquity to modern methods) parallels Kuhn's normal science phases, revealing overlooked features of incremental development in control frameworks.
- Implementing the concept of puzzles and problems in control systems reveals the achievements and engineering failures at various control systems paradigms and sets a logical transition from one paradigm to another.
- Applying the notions of conceptual discontinuities and incommensurability to the concepts and tools of sequential paradigms with modifications necessary for engineering application sheds light on the causes of engineering failures in control systems design.
- Kuhn's framework helps anticipate future paradigm shifts (e.g., data-driven control) and their implications for design methodologies.
- Examining control systems design through Kuhn's model clarifies the historical context in which specific design paradigms emerged and evolved. This contextualisation can provide valuable insights into past researchers' motivations, goals, limitations, and the factors that contributed to the development of current control systems paradigms.
- Studying the history and evolution of control systems design through Kuhn's model develops a deeper understanding of the underlying concepts, theories, and methodologies. This enhanced knowledge can lead to better critical thinking and problem-solving skills.

This study demonstrates how Kuhn's framework of scientific revolutions explains paradigm shifts in control system design. By analysing historical transitions through the Kuhnian approach, we develop a model to:

1. Explain past paradigm shifts,
2. Identify current anomalies driving change, and
3. Anticipate future transformations.

These technical and historical analyses validate and adapt Kuhn's core principles to engineering's unique epistemology, where knowledge advances through iterative problem-solving rather than purely theoretical falsification.

From a historical perspective, technical arguments concerning paradigm shifts within control systems design are emphasised. Kuhn's theories of scientific revolutions and paradigm shifts are utilised to develop a model for understanding these shifts in control systems design.

The historical timeline of control systems paradigms can be divided into the following six main periods, as discussed in Chaps. 2, 3, and 4:

- The pre-history and primitive control era: Control systems design was regarded as an art.
- The pre-classical period: Mathematics was introduced into systems stability analysis.
- The advent of system theory: Classical control was formalised, and early data-driven tuning techniques and mathematical system theory emerged.
- The emergence of modern control systems analysis and design: Rigorous mathematical framework dominated.
- Postmodernism: Uncertainty combat methodologies were developed to resolve anomalous results from modern model-based methodologies.
- The data-driven techniques re-emergence.

Paradigm shifts in control system design can be classified into three distinct overarching paradigms:

- Incorporating mathematics and physical laws into control systems analysis led to a transition from the artistic control design approach to a focus on mathematical descriptions.
- Introducing system-theoretic model-based control systems signifies a departure from the case-based designs of the pre-classical control era.
- Renaissance of data-driven control techniques.

However, the evolution of model-based control that represents a paradigm shift can be bifurcated into two distinct phases: the modern control era and the post-modern control era. The modern period is characterised by rigorous mathematical formalisation, featuring precise state-space formulations and deterministic optimal control frameworks. This phase assumed perfect model knowledge and focused on analytical optimisation. The subsequent postmodern era emerged in response to the fundamental limitations of these deterministic approaches, shifting focus toward uncertainty quantification and complexity management.

8.2 Scientific Revolutions and Paradigm Shifts

Thomas S. Kuhn, a prominent philosopher of science, revolutionised our understanding of scientific progress through his seminal 1962 book, *The Structure of Scientific Revolutions.* In his book, he introduced the concept of *paradigm shifts,* which is considered one of the most cited academic works in the social sciences [2, 3].

Kuhn's contribution is in proposing paradigm shifts in advancing scientific knowledge, challenging the idea of linear and continuous progress. As Kuhn defined it, a *paradigm* is a comprehensive framework encompassing various aspects of a scientific discipline during a specific period [3]. This framework includes established objectives, guiding principles, prevailing theories, methodologies employed for research, shared values and beliefs within the community, measurable indices for evaluating progress, and a set of guidelines and rules that direct scientific inquiry.

By introducing the concept of paradigm shifts and emphasising that scientific progress is not always a linear trajectory, Kuhn's work challenged the conventional view of science as a steady accumulation of knowledge. Instead, he highlighted the significance of revolutionary shifts in scientific thinking, which can lead to fundamental transformations in how we perceive and explore the world around us. This perspective has profoundly impacted the philosophy of science and our understanding of how scientific knowledge evolves over time.

In Kuhn's epistemological framework, the principal concept of a paradigm can be analytically substituted by two interdependent notions: the *disciplinary matrix* and *exemplars.* The disciplinary matrix constitutes the theoretical foundation of a scientific community, formally comprising (1) fundamental governing equations, (2) empirically validated reduced-order models, and (3) standardised performance metrics for design validation. Conversely, exemplars are benchmark problem solutions that demonstrate a practical application of theoretical principles and establish standardised methodologies. This dual structure explains how scientific traditions maintain coherence during normal science, where the matrix provides explicit theoretical rules, while exemplars offer implicit knowledge through standardised practice. The distinction proves particularly valuable when analysing control engineering paradigms, where matrix elements might include state-space formulations, and exemplars could entail benchmark control problems [4].

In Kuhn's framework, scientific progress occurs through alternating normal and revolutionary science phases. During *normal science,* which in [5] is described as "normal science is meant to bring out the idea that this type of science is business as usual", practitioners operate within an established paradigm, working to:

- Articulate and refine shared fundamental principles.
- Extend the paradigm's explanatory power through routine puzzle-solving.
- Preserve a common established technical tradition through standardised exemplars.

During the phase of normal science, new theories take shape and *evolve* within the established paradigm. *Revolutionary science*, by contrast, emerges when persistent anomalies destabilise the paradigm, necessitating a complete conceptual restructuring. These transitions are not merely incremental but represent fundamental transformations in:

- Core theoretical assumptions.
- Methodological approaches.
- Criteria for validating problems and solutions.

Exemplary instances in Kuhn's theory refer to specific cases, experiments, or methodologies that are particularly successful in demonstrating the effectiveness of a prevailing scientific paradigm. These instances showcase the paradigm's ability to explain phenomena and solve encountered challenges within its scope. They exemplify the paradigm's power and reinforce its legitimacy within the scientific community. We use the terms exemplary instances and exemplars interchangeably.

The concepts of *puzzles* and *problems* or *anomalies* play a pivotal role in distinguishing between normal science and revolutionary phases. Puzzles are well-defined challenges solvable within the paradigm's rules. In contrast, problems remain unsolvable and are persistent failures that can not be solved under current theories, regardless of the degree of creative work and effort invested to solve them.

Normal science is *puzzle-solving* and not *problem-solving*, as Kuhn defines it. In Kuhn's framework, scientific progress alternates between periods of normal science and revolutionary paradigm shifts. During normal science, researchers operate within an established paradigm, focusing on routine puzzle-solving activities that align with the paradigm's theoretical achievements and employ predictable solution strategies. However, when problems and persistent anomalies emerge and can not be solved, they gradually erode confidence in the existing framework. As these unresolved challenges accumulate, the scientific community reaches a crisis point where ad-hoc modifications are no longer sufficient. The encountered problems must be solved, and the growing frustration leads to revolutionary changes.

Kuhn's model of scientific change encompasses five phases, as follows:

1. *Pre-Paradigm Phase*, where competing, incomplete theories coexist and there is no consensus on fundamental principles. As this phase draws to a close, the scientific community gradually unites around a unified conceptual framework.
2. *Normal Science Phase*, where the community unifies under a dominant paradigm and focuses on puzzle-solving within established rules.
3. *Crisis Phase*, where accumulating anomalies arise, and the paradigm's solution toolkits cannot resolve them.
4. *Revolutionary Phase*, where radical reassessment of core assumptions starts, and a new paradigm starts to emerge to fundamentally alter the scientific landscape.
5. *Post-Revolution Normal Science*, where the community adopts the new paradigm and routine puzzle-solving resumes under the new and revised rules.

The paradigm is taken for granted during normal science as an unchallenged framework. Its theories are applied to solve discipline-specific puzzles. This conservatism enables cumulative progress but becomes limiting when problems and persistent anomalies emerge. These foundational assumptions undergo radical scrutiny in revolutionary phases, allowing revolutionists to resolve previously intractable problems that normal science could not address. The resulting paradigm shift redefines what constitutes valid problems, methods, and solutions.

In normal science, scientific progress is continuous, whereas *conceptual discontinuities* (representing a departure from the previously accepted framework to a new and often incompatible one) stand out as indicators of revolutions, leading to *incommensurability* within them.

Kuhn's concept of incommensurability captures the profound conceptual dislocation that occurs during paradigm shifts, where pre- and post-revolutionary frameworks become fundamentally untranslatable. They are no longer merely incompatible in their conclusions but mutually incoherent in their core definitions, methods, and validation standards. As Kuhn emphasised, the new normal-scientific tradition emerging from revolution is "not only incompatible but often actually incommensurable" with its predecessor, indicating that key terms, problem-solving tools, and criteria for acceptable solutions undergo radical redefinition [2].

8.3 Revolutions in Control Systems Design Through Kuhn's Perspective

Before investigating the technical aspects of Kuhn's paradigm shifts theory in control systems design in more detail, it is worth considering why this framework can matter for control engineers. The potential benefits of this study are outlined as follows:

- Paradigm shifts inspire a departure from established analysis and design norms and encourage innovative thinking. Applying this concept to control systems design could lead to exploring new methodologies, technologies, and approaches that might bring about significant advancements in the field.
- With the advancement of technologies and the emergence of complex and complex adaptive systems, existing control design paradigms encounter limitations or fail to address emerging challenges. By embracing paradigm shifts, control systems designers can proactively recognise these limitations and move towards new paradigms that better align with evolving needs.
- The current rapid evolution of technology can make existing control approaches obsolete. Paradigm shifts enable control systems designers to adapt to new technologies and incorporate them effectively, ensuring that control systems remain relevant and efficient.

- Complex and multidimensional problems often require a fresh perspective. Paradigm shifts encourage reevaluating fundamental assumptions and methodologies, potentially leading to breakthrough solutions for complex control-design-related challenges.
- Paradigm shifts often involve integrating insights from various disciplines. In control systems design, incorporating ideas from fields such as artificial intelligence, machine learning, and data science through paradigm shifts could lead to more comprehensive and practical solutions.
- Control theory education is primarily devoted to mathematical design techniques and simulation-based studies. Introducing Kuhn's ideas in control systems education could initiate and empower critical thinking and a deeper understanding of the evolution of control theories. Students would learn the current paradigm, the historical context, and the reasons behind paradigm shifts.
- Predicting future trends and challenges can be challenging. Control systems designers can be better prepared to face unforeseen disruptions and changes while analysing the anomalies of the settled control design paradigm.

From Kuhn's perspective, control engineers during normal science phases focus on puzzle-solving within established paradigms, designing for stability and performance under known theoretical constraints. This approach has yielded remarkable progress, from PID optimisation to robust control synthesis. However, our earlier discussion noted that accumulating anomalies eventually necessitate paradigm shifts.

The historical progression of automatic control advancements can be categorised into six primary periods. These periods, which will be further detailed in the subsequent sections, encompass the pre-history and primitive control era, the pre-classical period, the classical period, the modern control period, the postmodern period, and the resurgence of data-driven techniques.

While Kuhn's original paradigm concept faced criticism for ambiguity [6], its application to control systems design can be operationalised through engineering-specific criteria. In control systems design, paradigms are observed from the perspective of control design *practicability-utility*. A design school persists only if it successfully solves real-world problems. Furthermore, paradigm shifts occur in the case of accumulated *engineering failures* in a design school. For a definition of engineering failure, refer to Sect. 7.2.

Practicability-utility is a selected index or criterion to identify what constitutes normal science in control systems design. It is being used to assess control design methodologies' practical *feasibility*, *applicability*, and *workability*. By evaluating how well the paradigms align with real-world implementation and whether the design approaches are practical and usable, it can be identified whether a particular approach falls within the realm of normal science in control systems design. Note that practicability-utility must be the main characteristic of a design methodology, as control systems are utilised in each period to solve the proposed puzzles and problems.

While there may exist areas of overlap between various control systems design paradigms, defining distinct periods of normal control science becomes feasible by implementing a *feasibility-applicability-utility-workability index,* which can be equivalently regarded as a *non-failure index*. For a discussion on the multiple overlapping engineering theories, refer to Sect. 1.3. The control systems historical periods, marked by prevailing consensus, encompass a high degree of agreement concerning theoretical assumptions and the specific issues that these assumptions seek to address within the paradigm's framework. Note that assumptions in control systems design are equivalent to the notion of requirements addressed in Sect. 1.3.

Throughout the six distinct periods, control system design demonstrates a blend of practical engineering proficiency and foundational theory that capably resolves the challenges posed by control design puzzles. Even emerging anomalies hold insufficient weight to challenge the established techniques and theories in the early stages.

However, technological progress introduces control design problems of more complicated and complex systems. This leads to an escalating array of unresolved puzzles and anomalies, gradually fragmenting the confidence of control system designers working within the paradigm. As this confidence crisis deepens, compounded by the pressing demands of advancing technology, a stage is reached where a paradigm shift becomes imperative.

The degraded confidence and the compelling need for technological advances lead to the necessary derive for changes in control system design paradigms. The transition becomes not only a response to the theoretical challenges of control design science but also a development aligned with the necessities of the technological settings.

The terms *paradigm* and *paradigm shift* are often used loosely in control theory literature, frequently deviating from Kuhn's original philosophical construct. Many claimed paradigms merely describe methodological variations—such as fuzzy logic controllers being labelled a paradigm when they represent an implementation approach within existing frameworks [7] or computational intelligence techniques being framed as distinct paradigms despite sharing foundational assumptions with classical methods [8]. Even studies that apparently align with Kuhn's concept, like those proposing nonlinear model-free control strategies or attempting to bridge theory-practice gaps, typically introduce evolutionary refinements rather than revolutionary shifts [9, 10]. These innovations may expand the toolbox of normal science by addressing specific practical limitations, but they lack the conceptual incommensurability and theoretical separation that define genuine paradigm shifts. The conceptual blurring of incremental progress with true paradigm shifts suggests a need for a more rigorous application of Kuhn's criteria in control engineering discourse, particularly when distinguishing between advances that solve puzzles within established frameworks versus those that redefine them.

The keywords in Kuhn's concept of paradigm shift are *revolutionary science*, *exemplary instances*, *puzzles*, *normal science*, *anomalies*, and *incommensurability*. Most claimed paradigm shifts in control theory are evolutionary methodological advances, not true Kuhnian revolutions. Even notable cases, as the mentioned papers introduce incremental improvements within existing frameworks, lack the incommensurability and conceptual rupture of genuine paradigm shifts. These represent normal scientific progress, not disciplinary transformation.

The argument presented in this chapter asserts that paradigm shifts and scientific revolutions, as defined in Kuhn's concept, have taken place within broad frameworks of control theory. This will be further detailed in Sect. 8.4. During paradigm shifts, control scientists and engineers no longer share a consensus on control system design principles. This loss of trust in a design paradigm is evaluated with the feasibility-applicability-utility-workability or non-failure index. Control theory is useful if it can solve a practical, real-world application; otherwise, it remains a mere mathematical result.

The transition of control systems design from an individual-artistic approach to a more mathematically oriented approach institutionalised mathematics as the primary control analysis and design tool made many traditional heuristic methods *obsolete*.

The model-based control design paradigm made analysis and design heavily dependent on specific mathematical models and their underlying assumptions. Later, postmodern control techniques emerged to address uncertainties in these models and assumptions, forming a new class of methods to mitigate engineering failures.

However, in this transition, not all model-based techniques became obsolete. Instead, efforts were made to restructure design frameworks and develop new mathematical tools while preserving valuable aspects of earlier approaches.

Kuhn's description of normal science [2] appropriately captures the status of a control system designer during periods of a settled control design paradigm: "If it fails the test, only his own ability, not the corpus of current science, is impugned. In short, though tests frequently occur in normal science, these tests are of a peculiar sort, for, in the final analysis, it is the individual scientist rather than the current theory which is tested." This reflects how designers operate within established paradigms, where failures typically question their competence rather than the foundational theory itself.

Consistent with Kuhn's *five criteria* for evaluating scientific theories [2, 6], an effective control theory should exhibit the following qualities:

1. *Accuracy*—The theory must yield precise results within its domain. In control system design, this means that analytical and design techniques should be empirically validated, with measurable effects in both experimental and real-world implementations.
2. *Consistency*—The theory should produce coherent outcomes. When applied to a given system, its results should align with those derived from other methodologies within the same paradigm.
3. *Broad Applicability*—The theory should generalise well across diverse systems and remain robust under unanticipated conditions.
4. *Simplicity*—The theory should avoid unnecessary complexity. Its application, from simple to complex systems, should not require fundamentally different structures, costs, or assumptions beyond those associated with the observable data.
5. *Fruitfulness*—The theory should drive further discovery. In control design, this means either integrating advancements from other methodologies or adapting through theoretical refinements to improve performance.

Kuhn argues that these criteria form a *paradigm-neutral framework* for theory selection [6], enabling objective comparison between competing control methodologies.

8.4 Paradigm Shifts in Control Systems Design

The evolution of control theory can be divided into six key historical periods[2]:

1. *Pre-history and primitive control* (up to 1900),
2. *Pre-classical period* (1900–1935),
3. *Classical period and early data-driven methods* (1935–1950s),
4. *Modern control era* (1950s–1960s),
5. *Postmodern uncertainty mitigation* (1960s onward), and
6. *Resurgence of data-driven techniques* (past two decades to present).

In the subsequent sections, these transitions are modelled as paradigm shifts from a historical-philosophical standpoint.

8.4.1 *The First Paradigm Shift from Primitive Control to the Mathematical Age of Control Theory*

The origins of feedback control systems trace back to ancient innovations such as water clocks and the mechanical regulators described by Heron of Alexandria in the first century A.D. By the ninth century, explicit feedback principles appeared in early on–off controllers documented in *The Book of Ingenious Devices* and *The Book of Tricks* (~850). Historical records suggest that windmills were utilised in Persia as early as the seventh century for various tasks, including milling grain and drawing water from wells, often incorporating simple feedback mechanisms. Many of these elementary feedback mechanisms were found to be applicable during the onset of the Industrial Revolution. Refer to Chap. 2 for further details.

The era from antiquity to the pre-industrial period, as discussed in Sect. 2.2.1, can be viewed as the pre-paradigm stage of control systems design. During this time, the principles of feedback were initially established in simple artefacts such as water clocks by Egyptian and Greek communities of designers. More advanced schools emerged from ~800 to 1200 AD, leading to detailed technical publications that laid the foundation for designed artefacts and feedback mechanisms. The first paradigm of control systems design emerged from the post-industrial revolution to the pre-classical control era. Following the Industrial Revolution, a dominant methodology for control engineering was established, guiding the profession of control

[2] Note that the dates are rough and approximate, and they are not intended to be exact or precise chronological markers.

system design. Control systems were designed to regulate the performance of various systems across a wide range of engineering disciplines, as discussed in Sect. 2.3.

The first paradigm of control systems design was characterised by:

- Trial-and-error experimentation,
- Artistic/heuristic approaches, and
- Creativity-driven innovation.

This pre-mathematical era persisted until Maxwell's seminal 1868 analysis of the James Watt governor established the first mathematical framework for stability analysis. However, the systematic application of mathematical theory to control design only became widespread in the late 19th and early twentieth centuries, marking the transition to classical control theory [11].

The first paradigm shift in control theory marked the displacement of personalised, heuristic design methods by mathematical formalisation. This transition, which can rightfully be termed a *scientific revolution* in control theory, rendered prior techniques obsolete, replacing craft-based approaches with mathematically grounded frameworks.

Some researchers categorise these developing phases as the *pre-history* and *primitive* period of control systems design, preceding the emergence of systematic methodologies [12].

In the new paradigm, systems were modelled using the physical laws that governed their dynamics. This approach resulted in a set of differential equations that describe the system's behaviour. Mathematical analysis was introduced in the mid-1800s to analyse system stability. For the first time, designers applied physical laws to derive mathematical differential equations, analysed these equations and determined the parameter values for which closed-loop stability is guaranteed. This task was previously undertaken through trial and error and extensive and costly experimentation.

To analyse this paradigm shift through Kuhn's framework, we examine its key components: *exemplars*, *puzzles*, *normal science*, *anomalies*, and *non-competitiveness*.

Exemplars of the first paradigm shift are the combined mathematical analysis with heuristic methods and case-specific expertise to solve control design problems. The puzzles involved stabilising *simple systems*, including transportation (ship/aircraft navigation control), industrial processes, servomechanisms, and communication systems.

In Kuhn's words, the dominant design of that period, or normal science, was the *family of three-term controllers,* also known as *proportional-integral-derivative* (PID) controllers. As technologies advanced, anomalies emerged in the form of increasingly complex engineering systems that demanded the implementation of feedback control systems.

According to Kuhn's perspective, non-competitiveness refers to notions that became outdated in the preceding paradigm. During the first paradigm shift in control theory, artistic, heuristic, and individually-driven case-based design procedures that were once prevalent were replaced with new approaches. Applying mathematical analysis and implementing rigorous control systems design methodologies became the new paradigm's normal science.

Some notable anomalies where the established design approaches led to failures within the field of control systems design are as follows. As aircraft technology advanced, designers faced challenges in maintaining stable flight. Early aircraft were often unstable and required constant pilot intervention. This anomaly prompted the need for advanced control systems to stabilise aircraft automatically, and an autopilot should have been designed to adjust ailerons, elevators, and rudders automatically.

In chemical industries, the demand for consistent product quality and efficient operation led to the need for better process control. Traditional methods for controlling parameters like temperature, pressure, and flow often resulted in variations and inefficiencies. Engineers had to develop more sophisticated control strategies to ensure precise and reliable process control, addressing the anomaly of inconsistent product outcomes.

Anomalies emerged in the field of telecommunications as signal quality and strength became critical for effective communication. Traditional methods for regulating signal strength were inadequate to handle changing conditions and interference. Engineers had to devise control systems that dynamically adjusted signal strength and modulation to ensure reliable communication, especially over longer distances.

8.4.2 The Second Paradigm Shift to the Model-Based Design Paradigm

Mathematical models have revolutionised all engineering disciplines, becoming indispensable in research and design [13]. Model-based design has had a transformative impact on control systems, with both classical and modern approaches now relying on mathematical models in diverse forms. The second paradigm shift in control theory emerged by adopting system theory and model-based methodologies. This transition replaced heuristic, designer-dependent techniques (which used only elementary mathematical analysis) with comprehensive, model-driven frameworks grounded in general systems theory. For further details, refer to Chap. 3.

The primary characteristics of the established paradigm during the pre-classic control period of 1930–1935 were as follows:

- Simple mathematical models based on physical laws were introduced, enabling straightforward stability analyses for simple systems.
- Engineers in various disciplines conducted analyses and designed systems within their respective domains without collaborating with engineers from other fields. This was primarily due to the absence of a common language and platform for scientific and technological discussions.
- The design methods lacked proper formalisation to facilitate easy analysis and design processes.

Over time, anomalies began to emerge within the established paradigm. It is stated in [11] that "As applications multiplied, engineers became puzzled and confused, controllers that worked satisfactorily for one application, or one set of conditions, were unsatisfactory when applied to different systems or different conditions: problems arose when a change in one part of the system (process, controller, measuring system, or actuator) resulted in a change in the major time constant of that part. This frequently caused instability in what had previously been or seemed to have been a stable system."

Anomalies and challenges that emerged within the pre-classical control period can be summarised as follows. The controllers had limited applicability and worked well for specific conditions or systems, exhibiting instability or poor performance when applied to different systems or operating conditions. This could lead to unexpected outcomes and difficulties in achieving desired control.

There were no generalised design methods, and engineers struggled to generalise their design approaches across various systems. This could result in ad-hoc solutions that worked in specific cases but failed to provide consistent results across different scenarios. The engineers have failed to provide concise solutions to challenges as the systems get more complicated.

Also, with the increasing exchanges between systems of different disciplines, this lack of cross-disciplinary communication hindered the sharing of insights and best practices, making it challenging to solve complex control-related issues. Hence, the changing technological landscape outpaced the existing control methods, causing unforeseen problems and unsolved challenges.

Control scientists and engineers adopted *model-based design approaches* to address these anomalies. These models may be derived analytically from first principles, identified from data, or implemented as soft computing methods. Models and their role in control systems design are discussed in detail in Chap. 5. Model-based control design falls into three categories. The classical control approach, the state-space modern control approach, and the postmodern approach.

The mathematical tools and concepts relevant to the classical control approach included the Laplace transform for modelling and analysis, block diagrams graphical representations, frequency response analysis, mathematical performance specifications such as settling time, overshoot, and steady-state error, and applications of complex analysis in Nyquist-type stability and design methodologies.

Differential equations and linear algebra are the foundational mathematical tools extensively utilised in state-space modern control. Differential equations provide a means to capture and model the dynamic behaviour of systems, enabling the expression of relationships between states, inputs, and outputs. Linear algebra, on the other hand, empowers control engineers to represent these relationships in matrix form, facilitating the analysis, transformation, and manipulation of system dynamics. This combined utilisation of differential equations and linear algebra allows for the effective description, understanding, and design of state-space representations, enabling the implementation of advanced control strategies.

The second paradigm shift is the prevalence of models in control theory. However, as will be explained in Sect. 8.4.3, it is shown that this paradigm can be divided into two distinct approaches: the model-based control theory approach of classical and state-space modern control and the paradigm of postmodern uncertainty and complexity mitigation in control.

Although classical control and state-space modern control are two distinct approaches, state-space analysis can be considered an *evolution* of classical control. Indeed, many definitions and concepts in classical control were carried over to state-space analysis and design techniques. Key shared concepts include poles, zeros, and the subsequent analysis following pole-zeros configurations in dynamic systems, described using transfer functions or state-space equations. Performance metrics in both approaches, when devoid of uncertainties, exhibit similarities. Design techniques such as pole placement directly relate to both approaches, although using different representations. The choice between these approaches depends on the system's complexity, desired control performance, and the available tools and expertise.

In general, it can be stated that what can be achieved using the state-space approach is *attainable* within a transfer function framework, and vice versa, provided *exact* mathematical models are available.

The classical and modern model-based control paradigm can be systematically examined through Kuhn's framework of scientific revolutions, comprising three core components. Let us begin by considering the exemplars of the paradigm. The paradigm's exemplars consist of rigorous analytical methodologies rooted in mathematical system theory (Chap. 6), including transfer function-based input–output modelling, state-space system representations, and first-principles derivation of dynamical equations. These exemplars established standardised approaches for solving control problems through mathematical formalisation.

The puzzles within the model-based control paradigm were stabilising increasingly complex physical systems, achieving optimal performance under constraints, managing multivariable interactions, and addressing inherent nonlinearities. These challenges defined the research agenda during the paradigm's normal science phase. Also, the period of normal science was characterised by the dominance of frequency-domain analysis techniques, systematic state-space controller synthesis methods, optimisation frameworks, and algorithms for system identification. This methodological toolkit enabled engineers to solve industrial-scale control problems while operating within the paradigm's theoretical boundaries.

Anomalies emerged within the established model-based control paradigm as uncertainties emerged. These uncertainties violated the fundamental assumption of having an *exact* model. Researchers within the paradigm swiftly recognised that uncertainty had the potential to disrupt the core findings upon which the established normal science was built. Furthermore, the rigorous mathematical tools of mathematical systems theory were found to be insufficient for effectively addressing uncertainty.

To address the non-competitiveness issue of the exact model-based paradigm and the uncertainty combat paradigm that followed, we study the concepts or tools commonly used in deterministic analysis and design settings and become less applicable or even obsolete in the face of uncertainties. For further details, refer to Chap. 4. These include but are not limited to:

- Traditional optimisation and optimal control design techniques that assume precise and deterministic models and do not account for uncertainties.
- Techniques relying on exact linearisation around an operating point.
- Exact pole placement methodologies.
- Techniques that assume a known noise model for disturbance rejection, which do not hold in uncertain environments where the noise characteristics are uncertain or variable.
- Decoupling control systems to eliminate interactions between different control loops in a multivariable system.
- Deadbeat and minimum variance control approaches.

8.4.3 The Third Paradigm Shift to the Postmodern Era

By the early 1970s, unexpected failures in modern control implementations revealed fundamental limitations in the theory. Practising engineers frequently struggled with real-world applications, with model uncertainty identified as the primary cause. These challenges created a recognised *Theory-Practice Gap* in control engineering, which is discussed in Sect. 6.4.5.

While mathematical system theory remained central in control theory research, its perceived over-reliance on idealised assumptions caused the *holistic* view to be lost. Control scientists responded by developing different schools of thought, each addressing uncertainty through different philosophies.

After a decade of debate and research, three principal design philosophies emerged to form the *postmodern control paradigm.* Robust control (with fixed uncertainty bounds), Adaptive control (with time-varying parameter estimation), and Soft computing (AI-based approximation methods) based control.

Unified by their focus on *uncertainty mitigation*, these methods represented a global paradigm shift from modernism's *optimality-driven* approaches to postmodernism's *robustness-oriented* framework. Uncertainty and its different aspects and implications are discussed in Chaps. 4 and 5. This transition was driven by two key realisations: that exact mathematical models are seldom available in practice and that rigid model-based techniques frequently fail in industrial applications. Consequently, control theorists incorporated uncertainty quantification, addressed time-varying parameters, and developed alternative methodologies to traditional model-based design.

By the 1970s, control theorists and practising engineers increasingly challenged modern control theory's foundational dependence on precise mathematical models and their deductive frameworks. An important concern developed was that conventional methods prioritised abstract model manipulation over physical system behaviour, as examined in Sects. 3.3.5 and 6.4. These concerns fundamentally reshaped control philosophy, giving rise to three new methodological directions: *robustness*, *adaptation*, and *intelligence*. This conceptual shift solidified by the mid-1970s, stimulating the development of advanced uncertainty combat strategies that established new normal science for uncertain systems that displaced previous exact model-centric approaches.

Initial responses to these limitations emerged within the modern control paradigm as researchers sought to adapt existing methodologies to address uncertainty. Refer to Sect. 6.4.5 for more discussions on this, and in particular, refer to [14] for an early example. Some concepts from classical and modern control methodologies were upgraded and modified to conform with the new problems faced by control systems designers, as elaborated in Sect. 4.2.2. However, these incremental adaptations proved insufficient, and a broader philosophical shift was necessary.

Control theorists developed the new uncertainty combat paradigm through three key evolutionary steps: First, they formally incorporated uncertainty representations into system models. Second, they established rigorous frameworks for robust stability and performance guarantees. Finally, they systematically reformulated classical and modern methodologies to accommodate uncertainty in analysis and design procedures. This progression led to optimised robust stability and performance properties for feedback control systems. Also, researchers developed innovative tools within this restructured framework to extend modern control theory's capabilities [15, 16].

The adaptive control school emerged as a distinct philosophical approach to uncertainty management, complementing robust control methods. For more details on the necessity and emergence of adaptive control methodologies, refer to Sect. 4.3. Its core mechanism involves real-time identification of uncertain plant dynamics, automatic controller adjustment using estimated models, and application of the *certainty equivalence principle*, which is discussed in detail in Chap. 5.

While reducing the conservatism of robust control, adaptive approaches introduced new challenges such as stability and convergence issues, dependence on prior system assumptions, and the requirement for *persistent excitation conditions*. The persistent excitation constraint proved particularly limiting, necessitating input signals with sufficient richness to identify system parameters uniquely. It requires that the signal's richness corresponds to the number of parameters in the controller or the number of parameters in an adjustable model of the plant. These limitations prompted the adaptive control community to develop hybrid approaches (e.g., multiple-model adaptive control) and establish connections with robust and intelligent control theories. For more details, refer to Sect. 4.3.3.

Alternatively, and in a different theoretical and philosophical framework, uncertainty combat was followed by the soft computing-based control approaches. The soft computing-based control design approaches include three main methodologies: artificial neural networks, random optimisation techniques, and fuzzy logic. For more details, refer to Sect. 4.4. While traditional control theory derives its legitimacy from rigorous mathematical foundations, proved theorems, deterministic analysis, and exact solutions, soft computing methodologies operate on fundamentally different principles of approximate reasoning, heuristic optimisation, and data-driven adaptation [17].

Despite their differing philosophical foundations, adaptive control, robust control, and soft computing approaches share a fundamental model-based character that unifies them within the uncertainty combat paradigm. While their implementations vary, all three methodologies inherently depend on some form of mathematical modelling and require a priori system knowledge. Robust control relies on structured uncertainty bounds and nominal plant models, while adaptive control necessitates parameterisation assumptions and persistent excitation conditions. Similarly, soft computing methods, including fuzzy systems with linguistic rule bases, neural networks with architectural configurations, and stochastic optimisation within defined parameter spaces, all incorporate modelling assumptions. This common dependence means these approaches, despite their distinct uncertainty-handling strategies, all face the inherent limitations of model-based design: the need for system knowledge, structural assumptions, and the constraints these impose.

A comparison of the conventional model-based robust and (robust) adaptive control and intelligent controllers is provided in what follows.

It is argued in [17, 18] that the pragmatic approach embodied by neural networks and fuzzy logic methodologies seeks to address control problems directly, minimising explicit simplifying assumptions. However, given the inherent difficulty of real-world control problems, these methods ultimately yield approximate solutions to otherwise exact problem formulations. This characteristic reflects the fundamental trade-off between theoretical precision and practical applicability in soft computing approaches.

Whereas conventional model-based robust and adaptive control strategies seek exact solutions to approximate models derived via linearisation, affine parameterisations, and convex relaxations, intelligent control strategies pursue approximate solutions to realistic problems. This represents a fundamental philosophical inversion: *mathematical purity* versus *practical commitment.*

Within the uncertainty combat research community, an emerging solution framework is the hybrid methodologies that synergistically combine:

- The *analytical rigor* of conventional model-based approaches.
- The *practical adaptability* of intelligent control techniques.

This integration seeks to preserve mathematical tractability while accommodating real-world complexity, potentially mitigating the limitations of purely analytic or purely pragmatic paradigms. This will be further elaborated upon in the subsequent section of this chapter.

The exemplars of the postmodern uncertainty combat paradigm and its foundational tools comprise rigorous robustness frameworks, including uncertainty representations, and robust stability and performance, adaptation theoretical foundation, including stability and convergence guarantees for time-varying systems, artificial intelligence-based methodologies, including neural networks, fuzzy logic, and machine learning. The puzzles of the paradigm and key challenges include controlling, high-complexity systems, and uncertain dynamics, including aerospace, power generation, and process industries. The paradigm's normal science and the established methodology integrate robust and adaptive control designs, soft computing techniques, and hybrid analytic-pragmatic approaches.

Also, anomalies and persistent difficulties arise in complex adaptive systems, where traditional uncertainty-handling proves inadequate and safety-critical applications, where failures occur due to unanticipated interactions.

There are unresolved challenges across critical domains, revealing fundamental limitations in current control methodologies. Despite advanced decision algorithms, autonomous vehicles struggle with unpredictable human behaviour, sensor limitations, and ethical issues, as evidenced by accidents. Power grids remain vulnerable to cascading failures from component malfunctions, extreme weather disruptions, and cyber-physical threats, revealing gaps in large-scale adaptive control. Financial trading algorithms exhibit instability during market volatility due to unanticipated feedback loops and latency effects. In healthcare, ICU automation grapples with noisy physiological data interpretation and delayed adaptation to patient deterioration, leading to life-critical errors. Climate engineering proposals confront chaotic system interdependencies and irreversible threshold effects, highlighting the dangers of controlling incompletely modelled global systems. Even industrial process control, despite sophisticated control systems, suffers accidents from hidden process interactions, sensor degradation, and human factors. These persistent anomalies spanning transportation, energy, finance, medicine, environmental science, and manufacturing demonstrate the paradigm's unresolved difficulties in managing real-world complexity, where uncertainties compound across interconnected subsystems. For more details and societal challenges of control systems, refer to Sect. 10.9.

To address the non-competitiveness issue, the postmodern uncertainty-combat paradigm and the emerging data-driven paradigm, detailed in subsequent sections, selected key concepts from current uncertainty-based methods that face diminishing relevance in data-driven frameworks are presented:

- Uncertainty quantification, including structured and unstructured uncertainty representations and robust stability margins.
- Classical system properties, such as controllability and observability matrices and persistent excitation requirements.
- Classical stability and performance metrics and tools, such as Lyapunov-based stability proofs and frequency-domain performance metrics.
- Classical identification and optimisation frameworks.
- Stable and convergent adaptive strategies.
- Conventional intelligent control, such as first-generation fuzzy logic controllers and static neural network architectures.

8.4.4 The Paradigm Shift Towards Data-Driven Control Methodologies

Advanced model-based control techniques enhanced with robustness, adaptation, and intelligence have successfully addressed control problems for simple, complicated, and even some complex systems. However, as system complexity grows, so does the risk of control failure. The consequences of such failures become increasingly severe with system scale and complexity, where localised faults can cascade into large-scale disruptions. This relationship between complexity, risk, and failure impact is well-documented in domains ranging from power grids to aerospace systems, highlighting a fundamental challenge in modern control engineering. The control of future complex and complex adaptive systems is examined in [19], with the major societal challenges. For more details, refer to Sect. 10.9.

Complex and complex adaptive systems are expected to emerge more widely, exhibiting properties of complexity and adaptability. These systems have many interacting components that give rise to emergent behaviours and patterns. Complex adaptive systems are a subset of complex systems with the added characteristic of adaptability, meaning they can change and evolve in response to their environment or internal dynamics.

As technology advances and different domains become increasingly interconnected, the prevalence of future complex (adaptive) systems is expected to rise, resulting in unanticipated emergent behaviours. It is evident that the current control design methodologies within the established postmodern paradigm are insufficient to effectively design operational control systems capable of effectively managing these complex systems.

A defining characteristic of postmodern uncertainty-handling paradigms developed in recent decades is their inherent dependence on modelling assumptions required for theoretical derivations and mathematical proofs. As highlighted in [18], this reliance constitutes a fundamental vulnerability. When system behaviour deviates from these assumptions, the theoretical guarantees collapse. This tension between mathematical tractability and real-world complexity remains a central challenge in control design, as noted in [20]. The field continues to face a fundamental trade-off: either address the full complexity of engineering systems resulting in intractable *messy* problems or impose simplifying assumptions enabling elegant solutions at the cost of reduced practical relevance. Resolving this contrast to develop design tools that adequately address real-world control challenges remains an ongoing research frontier.

Within the model-based control paradigm, many researchers continue advancing existing techniques [21–23], maintaining that models remain essential for managing complex, interactive systems [21]. However, acknowledged limitations, including strong nonlinearities, unmodelable parameters, and stochastic complexities, necessitate complementing models with data-driven approaches. Specifically, systems with undefined mechanisms, massive variables, or heavy coupling increasingly require hybrid model-data solutions, particularly for autonomous applications.

A *pluralistic* philosophical framework for integrating diverse control methodologies to address emerging challenges is proposed in [17], advocating for methodological synthesis rather than paradigm competition. For further discussions on pluralism and its implications in engineering, particularly control systems design, refer to Sect. 9.5.

While a pluralist philosophy will undoubtedly enhance and further develop the existing control design methodologies within the established paradigm of model-based control, it cannot remedy the inherent limitations inherent in the *models* and *assumptions* of these techniques. This means that whenever models and assumptions are employed, their limitations are the foremost challenges in the practical control of complex adaptive systems.

In summary, the solutions put forth by control theory scientists within the current established model-based design paradigm are as follows:

- Further, develop and advance the existing techniques to address complex adaptive systems.
- Embrace a pluralistic perspective by integrating the present techniques with data-driven approaches.

However, it is arguable that the likelihood of success for the aforementioned proposals in real-world complex problems is limited. A pragmatic resolution to these challenges necessitates yet another *revolutionary paradigm shift*, one that surpasses the series of intricate assumptions. This shift requires a change in mindset, moving away from dependence on predefined constraints and prior knowledge and toward reliance on authentic field measurements. This signifies a genuine shift toward a data-driven control paradigm.

This new paradigm enables control systems to acquire operational data and dynamically reconfigure controller structures as needed. As foreseen by [22], the field has significantly transformed, indicating a major theoretical and methodological evolution. Over the past two decades, core elements of data-driven control have matured, yielding numerous fully data-centric design strategies that operate without reliance on first-principles models.

Despite significant theoretical advancements, the data-driven control paradigm has not yet achieved a state of normal science. Foundational innovations have emerged that are essential for this approach, including falsification frameworks for data validation, virtual signal architectures (encompassing error and reference signals), pseudo-model formulations, behavioural system paradigms, and rigorous data informativity metrics. The paradigm has also developed novel stability analysis methods, artificial intelligence and machine learning techniques, and explainable decision-making frameworks, all critical components for data-driven control systems. While simulations and laboratory experiments demonstrate promising results, real-world implementation of these methodologies remains limited in industrial applications. Before reaching paradigm maturity, the field must demonstrate practical solutions for controlling complex and complex adaptive systems, requiring further methodological refinement and documented success cases in operational environments.

A fundamental challenge in the data-driven paradigm involves establishing minimum data quality requirements analogous to persistent excitation conditions in adaptive control without relying on traditional system identification assumptions. The core objective is to derive effective controllers from datasets that may lack the richness and prior structural knowledge required for unique model identification. This necessity has encouraged the development of a rigorous framework for assessing *data informativity*: the study of necessary and sufficient conditions under which available data can guarantee specific control-theoretic properties when used for direct controller synthesis [24].

The new paradigm should overcome "*the three enemies of knowledge*," as acknowledged in [25]. The three enemies are *ignorance*, *uncertainty*, and *complexity*. Ignorance is the lack of awareness about something and not being aware of one's lack of knowledge [25].

Control engineering practice reveals a persistent form of *professional ignorance*: the necessity of learning from iterative design failures. Control system's historical devotion to the premise that *assumptions are prerequisites for knowledge* has been repeatedly contradicted by empirical evidence, particularly when systems encounter unmodeled dynamics or novel operating conditions. Despite this demonstrated fragility, the assumption paradigm remains entrenched in learning theory, estimation, and control research, a form of practical ignorance manifesting when designers develop learning systems incapable of generalising beyond their training assumptions. This cycle persists when controllers fail on real-world data that violate their underlying theoretical assumptions, revealing an inherent conflict between formal methodological constraints and real-world adaptive performance.

The new paradigm must address systems where critical model knowledge is absent, and designers recognise this. Rather than relying on assumed dynamics, its methodologies should derive control strategies directly from empirical input–output measurements. This philosophy directly challenges conventional practice, as emphasised in [22], which regards the belief that the system to be controlled is given a priori and is conceived as a *cardinal sin of automatic control.*

The primary challenge associated with complexity is its unpredictability, which sets it apart from complicated systems characterised by numerous components and interconnections. It is important to emphasise that no single efficient mathematical tool is universally effective, especially in high unpredictability and rapid changes. Present mathematical tools often have limitations and might require significant adaptations or trade-offs to address such complexities. Therefore, it is crucial to thoroughly evaluate the suitability of available tools for the specific context and recognise that certain systems' complexity can challenge their mathematical handling.

Figure 8.1 depicts a two-dimensional picture of paradigm shifts versus plant classifications. Finally, Fig. 8.2 provides a general overview of the developments and paradigm shifts in control theory based on Kuhn's perspective [26].

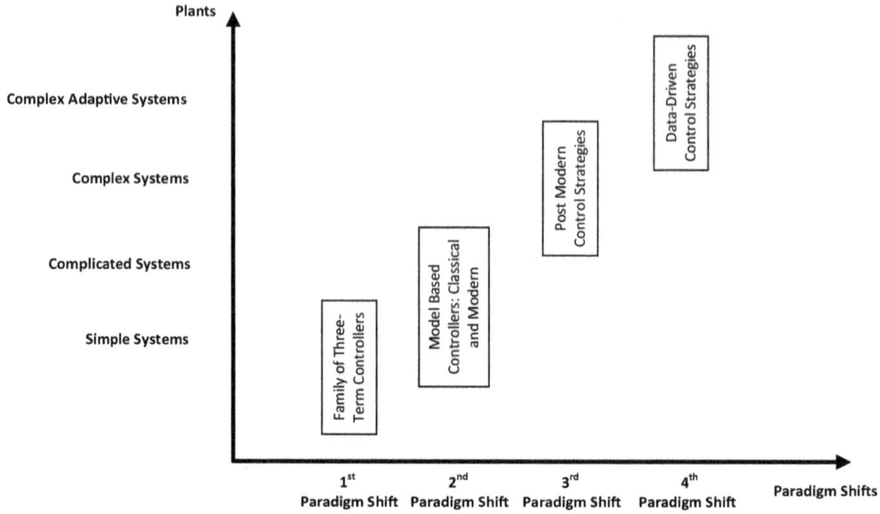

Fig. 8.1 A two-dimensional picture of paradigm shifts versus plant classifications

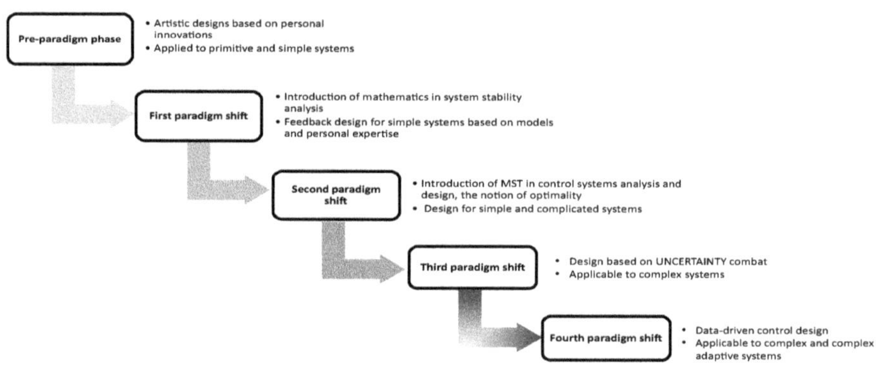

Fig. 8.2 A general overview of the paradigm shifts in control system design

8.5 Conclusions

This chapter analyses the evolution of control systems design utilising Kuhn's paradigm shift theory. Three historical paradigm shifts in automatic control are identified, with a fourth currently emerging to address the fundamental challenges of assumption dependence, uncertainty propagation, and adaptive complexity. Within established paradigms, design puzzles have been successfully resolved by prevailing theories. However, as technological advancements introduce new requirements and more complex systems, these frameworks increasingly produce anomalous results. The resulting anomalies have caused the paradigm shift toward data-driven methodologies, a direction actively explored by some control communities for over two decades. However, significant theoretical gaps persist regarding the epistemological foundations of data-driven control, formal stability guarantees without explicit models, and data quality and informativity metrics. Until these issues are resolved, the data-driven approach cannot mature into the normal science of a settled paradigm. This transition demands a rigorous examination of philosophical frameworks and mathematical foundations to establish a new paradigm capable of addressing modern control challenges.

References

1. P. Wernick, T. Hall, Can Thomas Kuhn's paradigms help us understand software engineering? Euro. J. Inf. Syst. **13**, 235–243 (2004)
2. T. Kuhn, *The Structure of Scientific Revolutions (1962)* (University of Chicago Press, Chicago, IL, 1996)
3. A. Bird, Thomas Kuhn, in *The Stanford Encyclopedia of Philosophy*, ed. by Edward N. Zalta (2018)
4. S. Psillos, *Philosophy of Science AZ* (Edinburgh University Press, 2007)
5. T. Lewens, *The Meaning of Science: an Introduction to the Philosophy of Science* (Hachette UK, 2016)
6. W.H. Newton-Smith, *The Rationality of Science* (Routledge, 2002)
7. F.L. Lewis, K. Liu, Towards a paradigm for fuzzy logic control. Automatica **32**(2), 167–181 (1996)
8. A. P. Engelbrecht, *Computational Intelligence: an Introduction* (Wiley, 2007)
9. Z. Gao, Y. Huang, J. Han, An alternative paradigm for control system design, in *Proceedings of the 40th IEEE Conference on Decision and Control (Cat. No. 01CH37228)*, vol. 5 (IEEE, 2001), pp. 4578–4585
10. Z. Gao, Active disturbance rejection control: a paradigm shift in feedback control system design, in *2006 American Control Conference* (IEEE, 2006), p. 7
11. S. Bennett, A brief history of automatic control. IEEE Control Syst. Mag. **16**(3), 17–25 (1996)
12. B. Friedland, *Control System Design: an Introduction to State-Space Methods* (Courier Corporation, 2012)
13. D.P. Michelfelder, N. Doorn, *The Routledge Handbook of the Philosophy of Engineering* (Routledge, 2020)
14. A. Fuller, Directions of research in control. Automatica **1**(4), 289–296 (1963)
15. P. Dorato, A historical review of robust control. IEEE Control Syst. Mag. **7**(2), 44–47 (1987)
16. M.G. Safonov, Origins of robust control: early history and future speculations. Annu. Rev. Control **36**(2), 173–181 (2012)
17. T. Samad, Postmodernism and intelligent control, in *Proceedings of 12th IEEE International Symposium on Intelligent Control* (IEEE, 1997), pp. 37–42
18. M.G. Safonov, Robust control: fooled by assumptions. Int. J. Robust Nonlinear Control **28**(12), 3667–3677 (2018)
19. A.M. Annaswamy, K.H. Johansson, G.J. Pappas,Control for societal-scale challenges: road map 2030. IEEE Control Syst. Soc. (2023)
20. H. Elmqvist, S. Mattsson, M. Otter, K. Åström, Modeling complex physical systems, in *Control of Complex Systems* (Springer, 2001), pp. 21–38
21. F. Lamnabhi-Lagarrigue et al., Systems & control for the future of humanity, research agenda: current and future roles, impact and grand challenges. Annu. Rev. Control **43**, 1–64 (2017)
22. K.J. Åström, P.R. Kumar, Control: a perspective. Automatica **50**(1), 3–43 (2014)
23. R.M. Murray, K.J. Astrom, S.P. Boyd, R.W. Brockett, G. Stein, Future directions in control in an information-rich world. IEEE Control Syst. Mag. **23**(2), 20–33 (2003)

24. H.J. Van Waarde, J. Eising, H.L. Trentelman, M.K. Camlibel, Data informativity: a new perspective on data-driven analysis and control. IEEE Trans. Autom. Control **65**(11), 4753–4768 (2020)
25. D.G. Elms, Achieving structural safety: theoretical considerations. Struct. Saf. **21**(4), 311–333 (1999)
26. A. Khaki-Sedigh, *An Introduction to Data-Driven Control Systems* (Wiley-IEEE Press, 2023)

Chapter 9
Postmodernism and Control Systems Design

Simplifying to the extreme, I define postmodern as incredulity toward metanarratives.
—Jean-François Lyotard

9.1 Introduction

The classification of control systems engineering is aligned with a historical perspective and is associated with different periods of the development of control systems technologies. Here is a breakdown of the proposed classification:

- Antiquity up to the Industrial Revolution
- Pre-Classical era
- Classical era
- Modernism
- Postmodernism.

The main characteristics of antiquity up to the Industrial Revolution era are the completely personal-heuristic-artistic nature of control systems designs with no apparent mathematical logic or structure. The gradual emergence of mathematics in the pre-classical control era is a turning point in control technology. Refer to Chaps. 2 and 8 for more details. The systematic study of linear time-invariant control systems with formal mathematics is the main feature of this era.

Modernism has had a profound effect on control systems design. Modernism in engineering is often marked by key characteristics such as a strong reliance on *rationalism*, emphasising knowledge gained through logical reasoning; *empiricism*, prioritising knowledge derived from practical experience; and *materialism*, focusing on a comprehensive understanding of the physical aspects of the universe.

Modernism has shaped the evolution of methodologies, technologies, and theoretical frameworks in control engineering systems. Key attributes of modernism in control engineering include a focus on mathematical formalism and optimality. Modernist principles underscore the importance of strong mathematical formalism,

A. Khaki-Sedigh, *Control Systems: A Historical and Philosophical Perspective*,
https://doi.org/10.1007/978-981-96-7426-8_9

incorporating advanced mathematical techniques and optimisation methods in the design and analysis of control systems. As indicated in [1], "control technology is often assumed by its practitioners to be a discipline that relies for its progress on advances in science and mathematics. Furthermore, this progress is taken to be a 'rational' process, in which theorems and other logical arguments furnish indubitable legitimacy to the new developments. The burden of mathematical proof distinguishes truth from falsehood (and the virtuous from the charlatans)." Another significant characteristic of modernism in control systems design is *determinism*, which tends to overlook uncertainties and assume exact information and conditions.

Postmodernism is a more recent intellectual and cultural movement that emerged in reaction to modernism. It gained prominence in the latter half of the twentieth century and is characterised by a departure from some of the core principles of modernist thought. Postmodernism can be seen as a reaction against the *certainties* and *grand narratives* of modernism. It challenges the idea that there is a single, objective truth or a universal way of understanding the world. In essence, postmodernism can be understood as a response to what it perceives as limitations or shortcomings in modernist thought. It questions the singular, linear narratives of progress and objective truth promoted by modernism, embracing a more pluralistic and relativistic view of reality. Postmodernism often emphasises subjectivity, cultural context, and the idea that multiple perspectives can coexist without one being universally correct. It is a shift from the confidence in reason, empirical observation, and a universal, reductionist framework, towards a recognition of diverse ways of knowing and interpreting the world.

Postmodernism is studied in the context of chemical and systems engineering [2, 3].

It can be argued that although *postmodernism* is challenging to define and encompasses various meanings that can differ based on the field and the intended purpose, there are shared elements among the different understandings of postmodernism. These shared elements can be identified, and their implications can be studied for a particular domain. One key intersection with postmodern thought in engineering involves ethical considerations, such as safety protocols, environmental pollution, and the prioritisation of human-centric values in technological decision-making, as studied in [2]. Ethical issues in control systems engineering have recently been raised. For more details, refer to Sect. 10.10.

The interplay of modernism and postmodernism becomes particularly evident in user-system interactions. Contemporary users increasingly demand systems accommodating diverse needs, cultural values, and evolving ethical standards. Requirements that often challenge traditional engineering assumptions about universal solutions. The discipline's capacity to integrate these postmodern critiques will largely determine its relevance in addressing the present challenges across various domains [3].

In this Chapter, we shall elaborate on the issue of postmodernism and control systems design from a control systems design context. As postmodernism is a diverse and multifaceted concept, it cannot be defined as a singular, unified idea or approach. This feature is exploited in control systems design.

9.2 From Modernism to Postmodernism

The term *postmodernism* first emerged in architectural discourse during the early 1960s as a critical response to the perceived limitations of modernist design principles. It subsequently became a label for a broader reaction against modernism across various fields [4]. Over time, postmodernism expanded beyond its original association with architecture. It evolved into a broader term describing a reaction to modernist principles in other fields such as literature, art, philosophy, culture, and engineering.

This expansion suggests that the dissatisfaction with modernism was not confined to one specific discipline. Postmodernism became a conceptual framework used to critique and challenge the assumptions, methodologies, and theories associated with modernist approaches across various domains. It is characterised by rejecting the strict principles of modernism, often embracing a more flexible approach to truth and representation.

The support for modernism initially stemmed from the mechanisation of industry, propelled by the invention of steam engines. Key features, including a rigorous reliance on mathematical formalism and a continual pursuit of abstraction and precision, characterised modernist science. Later, engineering advancements built upon these principles facilitated the exploration of alternative forms of energy and finer control over industrial processes.

Jean-François Lyotard, a prominent French philosopher, significantly shaped postmodern thought by challenging modernist narratives of progress. He characterized modernity as a period defined by grand narratives, which claimed universal truths and a prescribed course toward *utopian* ideals. In contrast, Lyotard defined postmodernity not as the culmination of modernity but as an era of skepticism toward these meta-narratives, where continuous change and fragmentation replace faith in absolute progress. This implies a fundamental shift in how we conceptualise system development, and the conventional notions of progress fail to account for the nonlinear complexities of contemporary technological and social systems. Lyotard further argued that traditional *master narratives* of progress, including positivist science, were no longer effective means of achieving advancement nor being relevant or successful in guiding society towards progress. In a postmodern condition, localised knowledge and pluralistic approaches replace these master narratives.

Master narratives are comprehensive, universalising claims that seek to explain and interpret developments in a field. These narratives operate as reductive epistemological frameworks for complex issues by framing them within a particular perspective. Key characteristics of master narratives include:

- *Universalism*, which often claims to be universally applicable, offering a *one-size-fits-all* explanation for diverse situations.
- *Totality* to provide a comprehensive understanding of a given subject, attempting to encompass all relevant aspects within a singular framework.
- *Legitimacy* and *acceptance* that make them authoritative.

In postmodernism, the concept of master narratives is often critiqued. Postmodernists argue that these grand, dominant narratives can be exclusive by not including or representing a wide range of perspectives and experiences, suppressing alternative perspectives and diverse voices. Instead, postmodernism encourages a more pluralistic approach that recognises the existence of multiple, sometimes conflicting narratives, rejecting the idea of a single, all-encompassing truth. An example of a grand or master narrative in control systems science is the linear systems theory advocated by Kalman and control modernists [1].

Hence, certain structuralist and modernist approaches, particularly those aimed at universal explanations or comprehensive narratives, were rejected in the coming outlook. Postmodernism, in contrast, tended to favour a more fragmented and pluralistic understanding, questioning the feasibility and desirability of all-encompassing narratives. This can be viewed as an overview of the differences between structuralist/modernist and postmodernist perspectives on explanation and narrative.

Postmodernism is aligned with the increasing complexity of our understanding of the world. The features of postmodernism that make this possible include tolerance of ambiguity, scepticism, mixing of styles, rejection of absolute truth, and lack of determinism. To elaborate further, postmodernism acknowledges and tolerates ambiguity and disorder, suggesting that the world is not easily reducible to clear-cut categories or fixed meanings. It also reflects a more critical and questioning stance. Finally, rejecting determinism and dogmatism suggests an openness to multiple perspectives and a reluctance to adhere rigidly to fixed doctrines [5].

While modern science primarily addressed matter and energy, postmodern science shifted its focus toward form and pattern [6]. Therefore, in science and, consequently, engineering, a distinguishing feature between modernism and postmodernism lies in their respective areas of emphasis. Modernist science has predominantly concentrated on analysing matter and energy, entailing an exploration of the physical properties, behaviours, and interactions of material substances and energy forms. In contrast, the postmodernist scientific perspective shifts its focus towards form and patterns. This signifies a departure from the exclusive concentration on matter and energy, leading to an increased emphasis on investigating structures, configurations, and recurring patterns across various phenomena.

9.3 Modernist and Postmodernist Control Systems Design Methodologies

It is difficult to draw a clear line between modern and postmodern control systems design methodologies. This section first provides a classification of modern and postmodern control systems design to tackle this issue. Then, it strengthens this classification by presenting and adapting it to the two metamorphoses of science.

9.3.1 *Classification of Modern and Postmodern Control Systems Designs*

The classification of modern and postmodern control systems lacks a precise definition. The few published papers on this subject refer to intelligent control as postmodern and consider the other methodologies based on classical mathematics as modernists. Reference [1] observes that emerging computational paradigms like neural networks, fuzzy logic, and evolutionary algorithms have often been regarded as interlopers-technologies with an intellectual heritage that puts them at odds with control science. With their distinct intellectual foundations, these technologies challenge conventional scientific approaches and have sometimes been dismissed as appealing primarily to non-specialists. The author further argues that intelligent system technologies reflect broader cultural shifts, emerging during an era when the modernist approach of rational reductionism faces growing skepticism across all domains of human inquiry, including science and mathematics. The debate between soft computing methods and mathematically rigorous control sciences reflects a deeper epistemological divide. This divergence stems from increasing skepticism toward rational/reductionist/modernist approaches, with intelligent system technologies embodying this critical shift in technological design philosophy. These approaches encompass all the other control systems design schools from modern control to robust and adaptive approaches. Intelligent control systems are therefore classified as the only postmodern control techniques. Also, in [7] all the control methodologies are considered classical to distinguish them from intelligent control systems and in [5, 6] fuzzy control systems are classified as postmodern.

This Chapter argues that modernism's key characteristics in control systems engineering include a strong emphasis on rationalism and determinism. These are the key characteristics of state-space and control systems designs based on the mathematical system theory, where uncertainty is not considered, and the mathematical models are taken as exact representations of the system.

For example, at the core of modern control, the observer-based state-feedback control systems design assumes that exact state-space matrices are available and that the corresponding characteristic equation provides the accurate eigenvalues of the system. These accurate eigenvalues are utilised to design a state feedback vector to place the closed-loop eigenvalues in other accurate locations. The same approach is applied in designing an observer gain vector for accurate observer pole locations based on the *separation principle.* This mathematically rigorous control system design with complete stability proof and guarantee of observation error convergence to zero can falter in the face of small uncertainties.

A new class of control systems has emerged to address uncertainty, as discussed in Chap. 5. These approaches deviate from strict rationalism and determinism in their design philosophy. They range from fixed structure robust controllers to nonlinear time-varying adaptive controllers, maintaining conventional mathematical rigor and nonlinear time-varying intelligent control with underlying soft computing principles.

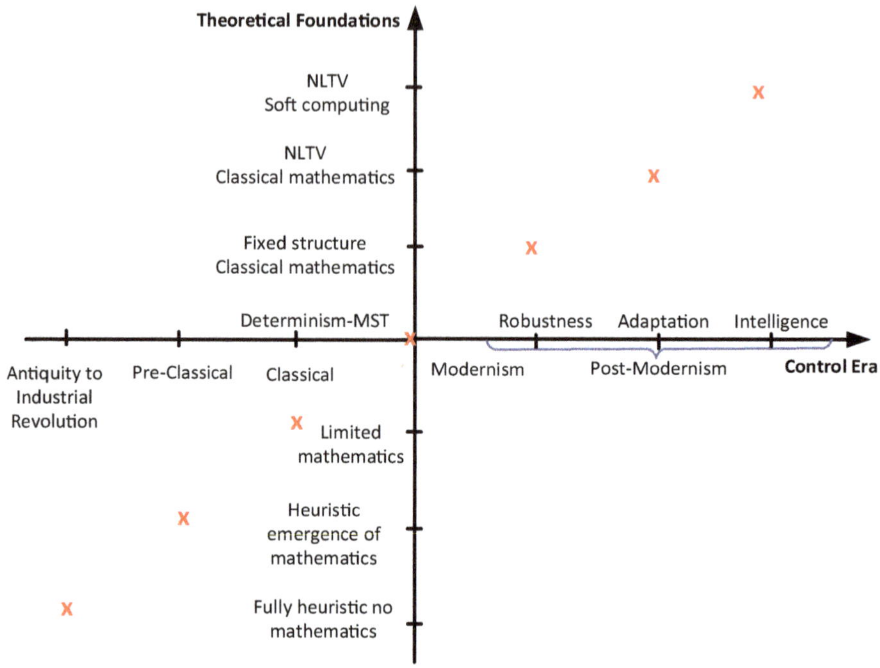

Fig. 9.1 Antiquity to postmodernism era of control systems versus their theoretical foundation

This spectrum of postmodern control design methodologies is a platform for controlling uncertain, time-varying, and unknown systems with ill-defined mathematical models.

This is depicted in Fig. 9.1, where controllers from the antiquity era to postmodernism are shown on the horizontal axis. The vertical depicts the range of controllers from fully heuristic with no theoretical or mathematical foundations to intelligent controllers.

9.3.2 *Modernist and Postmodernist Control System Designs Versus Metamorphoses of Science*

The classification of modern and postmodernism in control system design methodologies presented in Sect. 9.3.1 aligns with the analysis given in [8] regarding science's first and second metamorphoses. Projecting these metamorphoses onto control system designs demonstrates notable coherence with the aforementioned classification of modern and postmodernism in control system design methodologies.

The first metamorphosis of science transformed the foundational principles from those inherited from the time of Aristotle to a more modern form. The new operational foundations included the use of physical experiments and mathematical models, particularly involving differential equations. Along with these operational methods, certain philosophical attitudes and beliefs were developed. There was a change in mindset regarding the limitations of encoding and decoding information between experiments and mathematics. The concern about the accuracy of this information exchange diminished, leading to an increased identification of the physical world with mathematical models. This resulted in a growing tendency to identify the physical world with the mathematical models used for making predictions, even though these predictions were very limited. Philosophical beliefs emerged regarding reductionistic methods in science. This included the idea that it is possible to synthesise reductionistic knowledge to create a comprehensive theory of the systems capable of predicting all natural and man-made phenomena.

Over the last century, there has been a profound shift in the nature and structure of science through a second metamorphosis driven by two categories of discoveries. The first set of discoveries revolves around mathematical results directly tied to our limitations in understanding the dynamic behaviours in nature through mathematical models. Although these discoveries should have ushered in an era marked by a *loss of innocence*, indicating an acknowledgement of our limitations, they have largely escaped the notice of scientists. Despite being somewhat overlooked, these limitations carry significant implications for the future character of science. These discoveries, applicable to virtually all mathematical models of physical systems, underscore our inherent constraints in making (1) analytic mathematical deductions, (2) deterministic physical predictions, and (3) structurally stable models[1] of closed systems.[2]

In the latter half of the twentieth century, scientific investigations underwent a profound transformation with the advent of digital computers. This shift expanded the operational bases previously established during the first metamorphosis, namely *Physical Experiments* and *Mathematical Models*, to introduce a third crucial element: *Computer Simulations*. These computer simulations represented a new operational basis for acquiring knowledge about systems. The operations conducted within these three areas—physical experiments, mathematical models, and computer simulations—can independently contribute valuable insights into our understanding of systems. An important point is that computer simulations played a pivotal role in addressing and extending our understanding of key concepts. Specifically, the limitations or challenges faced when trying to derive conclusions or predictions from mathematical models can be clarified by providing visualisations and simulations that may not be easily achievable through mathematical models alone. Also, physical experiments often generate data that needs to be translated into mathematical terms for further analysis. Similarly, mathematical solutions sometimes need to be

---

[1] A structurally stable model is one that maintains its qualitative behavior under small perturbations or changes in its parameters. In the context of dynamic systems, it implies that the essential features of the system remain unchanged when there are small variations in its structure or parameters.

[2] Closed systems refer to systems that do not exchange matter with their surroundings. In a mathematical context, it often implies a self-contained and isolated system with well-defined boundaries. Refer to Sect. 6.2 for more discussions.

interpreted back into physical terms for practical applications. Computer simulations can bridge this gap by encoding physical experiment data into mathematical models and decoding mathematical solutions into physical terms, thus improving the communication and understanding between these two approaches. The other role of computer simulations can be in the bifurcation processes of dynamic models, that is, splitting a system's behaviour into multiple possible outcomes as a result of changes in parameters. In dynamic models, these processes can be complex and difficult to analyse using only mathematical models or physical experiments. Computer simulations can help visualise and analyse these bifurcation processes more effectively, better understanding the system's behaviour under varying conditions.

For any practising engineer, it is a well-established fact that determining the state of a system comes with inherent limitations in accuracy. It is recognised that, in reality, we cannot realistically expect to reduce this uncertainty to arbitrarily small levels. This pragmatic viewpoint is not always as readily accepted in theoretical scientific circles, especially when dealing with highly complex physical situations beyond typical engineering scenarios.

Theoretical scientists sometimes hold a more optimistic and idealistic perspective regarding their ability to define conditions for physical systems precisely. What is not widely acknowledged is the significant impact this loss of accuracy has on our ability to predict future outcomes. Importantly, this lack of determinism does not affect all systems equally; certain systems remain deterministic over specific time scales [8]. The projection of these metamorphoses in the control systems design methodologies is illustrated in Fig. 9.2.

9.4 Postmodernism Implications in Control Systems Design

As previously discussed, postmodernism has led science and engineering towards a new fundamental vision of systems, with forms and structures alongside matter and energy that have been fundamental in modernism. The information about a system includes details about its form and structure, leading to the triad of *matter-energy-information* [5]. In control engineering, information, in the context of systems, can be broadly defined as data that has been processed or organised to be meaningful and useful. It represents knowledge or patterns derived from data. Whereas data refers to raw, unprocessed facts or observations. Refer to Sect. 7.3.2 for more elaborations. It becomes information when it is processed, interpreted, and organised to convey meaning or it can be directly used for control systems design as in data-driven control methodologies. When data from a system is processed and organised, it provides insights into the arrangement, patterns, and relationships within the system. Engineers and designers often gather data from a system, analyse it, and derive information that guides the creation of a specific form or structure. This quote attributed to Carly Fiorina, the former Chief Executive Officer of Hewlett-Packard, "*The goal is to turn data into information, and information into insight*," encapsulates the above

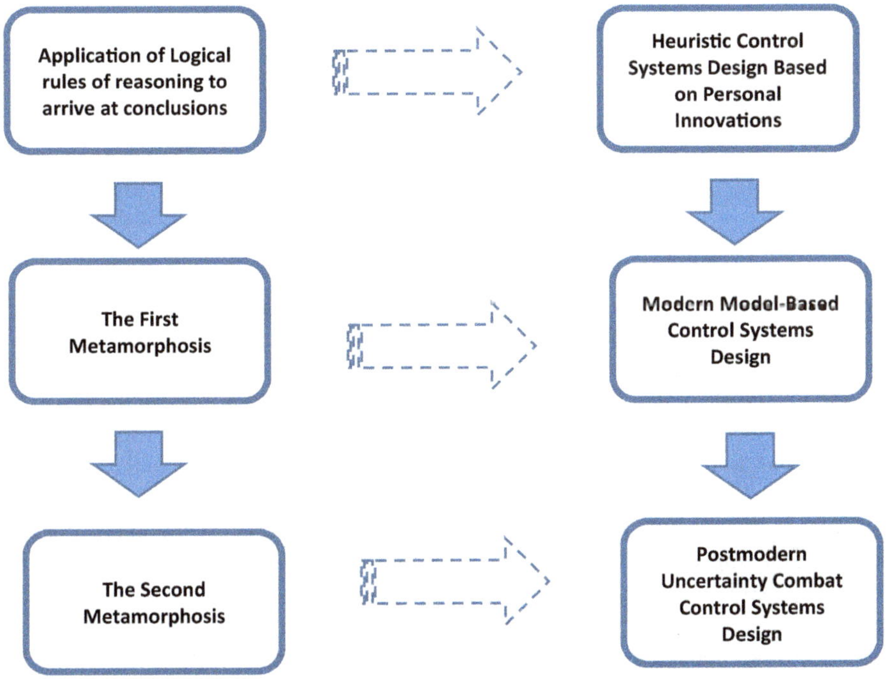

Fig. 9.2 The correspondence of control systems design and metamorphoses in science

process. For example, an adaptive design involves collecting data to determine a particular structure for the system or to design a controller.

In control systems engineering, the triad of matter—energy—information may be replaced by *physical systems-dynamics-data.* Physical systems encompass the objects or systems that are subject to control and manipulation. Dynamics represent the behaviour, motion, and changes within the physical systems. It includes the response of systems to various inputs and disturbances, reflecting the dynamic nature of control systems. Data represents the information and knowledge derived from the physical systems and their dynamics. It includes measurements, observations, and other relevant information for analysis and control. This is depicted in Fig. 9.3. It is evident that data/information plays a key role in the design of postmodern control systems.

Although the term postmodernism was first introduced in architecture in the 1960s, it was not until the 1970s and 1980s that postmodernism gained widespread recognition as a distinct intellectual and cultural movement. This coincides with the emergence of postmodern control systems design schools of robustness-adaptive-intelligence. Refer to Chap. 4 and Sect. 5.7 for more details.

In the field of industrial processes, the instrumentation of information is dynamically governed through robust-adaptive-intelligent control mechanisms. Either separately or in a hybrid design with a pluralistic approach, which will be discussed in

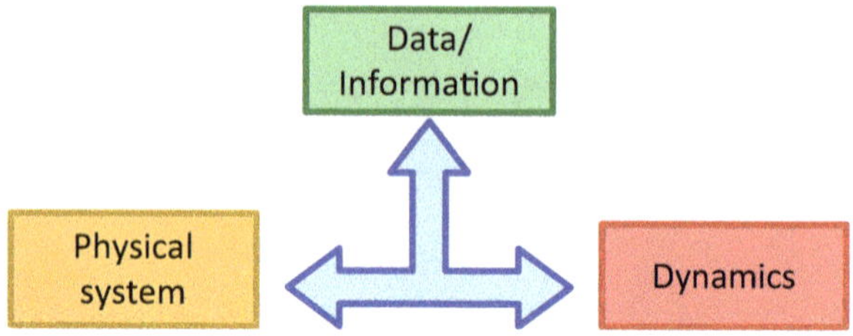

Fig. 9.3 The physical systems—dynamics—data triad

Sect. 9.5. This holistic approach recognises that controlling complex systems requires using data for a robust-adaptive-intelligent framework to changing conditions and withstanding uncertainties.

The postmodern control systems utilise information obtained from the physical systems they control. This information includes real-time data, measurements, and observations that provide insights into the current state and dynamics of the industrial processes.

Robustness, adaptation, and intelligence are pivotal aspects involving continuously adjusting control parameters in response to evolving conditions. With its learning-adaptive capabilities or robust nature, postmodern control ensures that the system dynamically responds to variations, disturbances, and uncertainties. This robustness-adaptability-intelligence is crucial for optimising performance and maintaining efficiency under changing circumstances.

Note that robustness is the first on the list of postmodern control strategies. Although robust control strategies have generally fixed structures, they are employed to fortify the system against uncertainties, modelling errors, and external disturbances. Rather than relying solely on dynamic adjustments, robust control focuses on preventive measures during the design phase, giving controllers that inherently exhibit stability and performance robustness across a range of uncertainties.

Consider the following example of the effect of switching controllers from [5]. A switching system's potential destabilisation through an appropriately chosen switching signal, even when transitioning between several closed-loop stable systems, can lead to catastrophic outcomes in systems employing multiple controllers and switching logic. Note that stability depends on initial conditions and this rejects the universal stability guarantees. Conventional system theory, relying on frequency analysis and the transfer function, fails to explain this effect due to its definition of null initial conditions, contrasting with real-world applications that typically have non-null initial conditions. The study of systems with null initial conditions simplifies equation manipulation, revealing characteristic behaviours aiding in system comparisons and constructing a general theory of systems. However, this quest for generalisation may result in unexpected failures in specific conditions. Operational calculus (Laplace) offers a comprehensive view of linear systems, establishing conventional linear PID control aligning with the modernist vision of universality and coherence. For nonlinear systems, frequency analysis struggles due to problem diversification and the absence of a unified theory. For barely controllable systems, the heuristic approach is the only unified method for now. Despite lacking rigour, heuristic solutions are flexible and comprehensive, offering positive results in various applications. Time analysis, facilitated by phase trajectory, provides valuable support for heuristic control decisions in applications, although not revealing much about internal system structures. In contrast to the Cartesian rigorous modernist vision, the heuristic approach aligns more with postmodernism. While not mathematically proven, heuristics consistently yield positive results in many applications. The traditional mathematical approach of hypothesis → conclusion → demonstration is transitioning to a more pragmatic methodology in the postmodern era: hypothesis → conclusion → computer simulation/experimental tests. Rather than solving differential equations manually, computers numerically integrate them. Consequently, most industrial products in the postmodern era are designed using dedicated software embedding domain-specific knowledge, often acquired through simulations and linguistically represented by expert systems. Rigorous experimental tests to validate the designs compensate for the lack of a rigorous theory.

9.5 Pluralism in the Postmodern Control Design Methodologies

Postmodernism provides a platform for a *pluralistic* approach to control systems design. Researchers have advocated this approach to overcome some problems associated with model-based control techniques and enhance closed-loop system stability and performance. This section provides a brief overview of postmodernism and pluralism from a general philosophical perspective. Then, the implications of postmodernism and pluralism are discussed in a general engineering context. Finally, postmodernism and pluralism in control systems design are critically reviewed.

Postmodernism and pluralism are two distinct but interconnected concepts that have significantly impacted various intellectual and cultural domains, including philosophy, art, literature, and social sciences. As previously noted, postmodernism is a philosophical and cultural movement that emerged in response to modernism's perceived limitations and failures. It challenges the foundational principles of modern thought, such as grand narratives, absolute truths, and fixed meanings. The key ideas of postmodernism can be summarised as follows:

- Postmodernism rejects overarching metanarratives. Instead, it emphasises local, diverse, and fragmented narratives.
- Postmodernism is sceptical about the possibility of objective truth. It argues that truth is socially constructed and varies across different contexts and perspectives.
- It involves analysing and dismantling existing structures to reveal their underlying assumptions and contradictions.
- Postmodernism embraces diversity and pluralism.

Also, pluralism, in a broader sense, refers to recognising and accepting diversity, multiplicity, and the coexistence of different theories and perspectives within a system. The key ideas of pluralism can be summarised as follows:

- Pluralism advocates for the acknowledgement and celebration of diverse theories.
- Pluralism suggests that there can be multiple valid approaches to functional problems. It opposes *monism*, which asserts that there is a single fundamental solution to such problems.
- Epistemic pluralism acknowledges multiple ways of knowing and understanding a system. It rejects the idea that there is only one privileged form of knowledge.
- Value (cost) pluralism contends that there are irreducible and even conflicting values (costs) that may not be hierarchically ranked in system designs.

Hence, it can be concluded that both postmodernism and pluralism challenge the notion of a singular, universal design perspective. The pluralistic nature of postmodern thought allows for the coexistence of various designs and values (costs) without imposing a hierarchical structure.

The philosophical ideas of postmodernism and pluralism can have implications in engineering, particularly in how engineers approach their work, interact with diverse stakeholders, and address complex challenges. Here are some potential implications:

- Diverse perspectives in design, where engineers may consider a broader range of perspectives, experiences, and cultural contexts in the design process. This can lead to more inclusive and user-friendly products and systems.
- Openness to multiple solutions by embracing pluralistic views in engineering fosters a recognition that there may be multiple valid solutions to a problem. This openness can encourage creativity and innovation by exploring various approaches.
- Interdisciplinary collaborations can be promoted by breaking down traditional disciplinary boundaries. Engineers may work more closely with professionals from diverse fields, bringing different expertise to solve complex problems.
- User-centered designs where the end-users' diverse needs and experiences are central to the development process.
- Adaptability and flexibility are encouraged in the engineering approaches.

The philosophical concept of the amalgamation of different control design methodologies to encounter new challenges or pluralism in control theory was raised in [1] and is practised by many researchers and control systems designers. In the specific domain of control systems design, pluralism refers to the *integration* of diverse methodologies, strategies, and viewpoints to develop effective control solutions for complex and dynamic systems. It recognises that there is no one-size-fits-all approach to control system design and that different techniques and approaches can be combined to achieve the best outcomes. Pluralism in control systems design acknowledges the complexity of modern systems and the need to adapt and innovate to address intricate challenges.

Quoting from [9], it is stated that promoting pluralism in science leads to a deliberate strategy of devising multiple competing theories to enhance testability. While a pluralist philosophy will certainly enhance and further develop the existing control design methodologies within the established paradigm of model-based control. However, it is argued that pluralism within the model-based school cannot remedy the inherent limitations of the *models* and *assumptions* underlying these techniques.

This means that whenever models and assumptions are employed, their limitations take priority as the foremost challenges in the practical control of complex adaptive systems. While pluralism can help overcome certain limitations, it is important to note that it introduces challenges, such as *increased complexity* and *potential integration issues*.

The potential integration issues in pluralistic control systems design, where diverse methodologies and approaches are integrated, can arise due to the complexity of combining different strategies. These issues can impact the effectiveness and reliability of the control system. Let us elaborate on some of these potential integration issues.

Different control strategies might have diverse requirements and assumptions that make them incompatible when integrated. This can arise in the integration of model-based designs and data-driven methodologies. Where assumptions and simplifications required for the model-based design can deteriorate the overall performance of a full data-driven control technique. This can also be viewed as interference with different methodologies, leading to unexpected behaviours or reduced performance.

Integrating multiple approaches can increase system complexity, making managing and troubleshooting the control system challenging. Different control strategies have different tuning parameters. Tuning these parameters for optimal performance is often complex, challenging, and time-consuming. Integrating different methodologies can potentially further escalate this problem.

Combining different control strategies can introduce stability issues, where the overall system becomes less stable than individual components. Validating the stability and performance of an integrated system becomes more complex. Identifying the root cause of stability and performance failure or malfunctioning could become more complex. Also, an adjustment or change in one system parameter might have unexpected consequences on other parts due to the interconnected nature of the strategies.

9.6 Internal Realism and Pluralism in Control Systems: A New Perspective on Reality and Control

In a model-based control process, the model captures, processes, and internalises knowledge about the system to be controlled, using this knowledge to guide the controller's actions based on its predictions. Such designs are either explicitly or implicitly based on models that attempt to encapsulate the knowledge available about the aspect of reality to be controlled.

In practical real-world systems, different conceptual modelling schemes and languages yield multiple models, rendering a single, unique model to represent the reality of a system impossible. This leads to different control design methodologies.

The system designer has many modelling languages and design techniques at his disposal including classic differential equations, discrete-time dynamical systems, discrete event systems, stochastic systems, intelligent and machine learning systems, and qualitative descriptions. These modelling tools give rise to different descriptions and design techniques for a given reality or system. The languages, representational systems or conceptual schemes attempt to describe the system elements and the existing interconnections between them.

From the first theoretical analysis of control systems, see Sect. 2.4, the modelling and design of control systems have elegantly fitted the classical areas of mathematics. A key assumption is that the controlled systems are governed by laws that are often deterministic and that classical mathematics would adequately represent their structures and behaviours. Even if other formalisms, such as difference equations, discrete-event, and AI-based models, are provisionally employed, the model given by differential equations has been most widely used and traditionally considered the right one. This is observed because most control theory is based on differential equations as an adequate formalism for the system description. Also, in the control engineering practice, this language is widely accepted as the true one to describe reality or, at least, the best one to do so. On the other hand, even with this formalism, different descriptions of the same system are derived. For instance, internal state space and external black box descriptions will lead to alternate mathematical models and design techniques.

While different mathematical formalisms may be used to describe control systems, the problem of alternate descriptions is not typically considered a significant concern by the control engineering community. They generally assume that these alternate models are simplifications of a more fundamental, underlying model. In other words, they assume that different formalisms can be reduced to a single, unified model, sometimes called a *base model*. This perspective is often implicit and is not necessarily considered a problematic or serious issue, as long as it is assumed that the alternate models are simplifications of a real model.

However, there are several serious objections to this perspective. The base model concept is a very strong and implausible metaphysical assumption, and the traditional approach relies heavily on this unifying assumption, which is problematic. The increasing complexity of hybrid modelling and design techniques, which combine different formalisms, highlights the limitations of this assumption. These hybrid approaches often involve formalisms that are not easily reducible to one another, let alone to the classical frameworks of differential equations and dynamical systems.

As a result, the traditional approach to modelling and designing control systems may not be sufficient for dealing with these more complex and non-classical models. Rather than dismissing these models as lacking a fundamental understanding, it is essential to re-examine the underlying assumptions of control systems modelling and design and consider alternative approaches that can accommodate the diversity of formalisms involved.

Control systems based on AI, machine learning, and data-driven methodologies are spreading, and a paradigm shift is occurring in the control systems realm. Refer to Chap. 8 for more details. One may contemplate that these approaches are relevant only because no more precise mathematical descriptions are available or due to the uncertainty of the information regarding the controlled system, and non-classic descriptions would be provisional until the classic standard ones could be found or more complex and detailed mathematical models and design techniques are derived based on classical mathematics.

However, the objection to this view is that the current paradigm's control systems modelling and design approaches are not reducible to models based on differential equations. They are preferable to the settled paradigm as they can solve control problems that the methodologies of the settled paradigm are unable to solve. Hence, we are faced with an unavoidable problem of pluralism. The control system has a concrete reality to control, and it needs to represent it somehow to control the physical system. This is an ontological aspect of control systems, which emphasises the need for a deep understanding of the physical reality being controlled. The control system needs to be able to reason about the world to control it effectively, and this reasoning is not just a matter of mathematical equations and algorithms but also requires an understanding of the physical and semantic meanings of the data being processed.

The fact is that there are always alternate descriptions, models, and methodologies that can be utilised. These methodologies will lead to a different controller design modelling and framework. A key question to be asked is: How can the designer decide which one of these different approaches is the most suitable?

At the heart of this question lies the fact that the designer is searching for an approach that best copies reality such as *it is in itself*, and derives the most appropriate controller. Here are where proposals for the base model have been made. But what exactly is copying reality as it is in itself? In what sense would the approaches be able to copy reality in every relevant aspect? As a matter of fact, the only thing that can be claimed is that the approach is appropriate if the controller based on it, gives rise to a closed-loop system that behaves in the way it is expected to. So, the question of whether the proposed approach manages or not to copy reality as it is in itself remains open.

There has been an interesting notion known as the *internal realism*[3] (IR) of Hilary Putnam[4] about realism in epistemology and philosophy of science, which could provide practical insights for the posed questions in control systems. It can be utilised to make clear the sense in which some kind of *pluralism* could be justified.

According to the concept of IR, the idea that there exists a single, definitive description or model that accurately captures the essence of reality is a *myth* [10]. In other words, IR suggests that there is no one true or objective description of

[3] This is also called by Putnam *pragmatic realism*, or simply *internalism*.

[4] Hilary Whitehall Putnam, an American philosopher, mathematician, and computer scientist, played a prominent role in the field of analytic philosophy during the latter half of the twentieth century. He made significant contributions to the study of philosophy of mind, philosophy of language, philosophy of mathematics, and philosophy of science.

reality that awaits discovery. Instead, our descriptions, theories, and models are all necessarily embedded within our conceptual frameworks and shaped by our unique perspectives and experiences.

This perspective challenges the traditional notion of *objectivity* in science, which aims to uncover an objective truth about the world. According to IR, our understanding of reality is always filtered through our own *subjective* experiences and *conceptual* frameworks, and it is impossible to know reality independently of these frameworks.

In the context of control systems, this means that there is no single objective truth about how a system should be controlled or what constitutes optimal behaviour. Instead, our understanding of a system is always shaped by our conceptual frameworks and goals, and different control strategies may be equally effective depending on the specific context.

IR also suggests that there may be multiple, equally valid conceptual schemes for understanding and controlling complex systems. This is because different perspectives and experiences can lead to different insights and understandings, and it is not necessarily possible to reconcile these differences or identify a single correct approach.

In practice, this means that control systems may be designed to accommodate multiple perspectives and goals and to be flexible and adaptive in response to changing circumstances. It also means that control systems may require ongoing monitoring and adjustment to ensure that they remain effective and efficient over time.

IR rejects the notion of a *ready-made* and *fixed structure* of reality where all our descriptions, theories, and models must perfectly match, a viewpoint known as *metaphysical realism* (MR) or *externalism*, as called by Putnam. According to MR, there would exist one true description, theory, or model for the whole of reality and for each of its parts, and the truth is defined by how well our descriptions align with the fixed structure of reality.

According to the perspective of MR, knowledge is seen as the process of uncovering the inherent structures of reality without being influenced by our understanding or contributions. Additionally, in the long run, the effectiveness of control models would rely on having this type of knowledge about the fixed structures of reality.

MR is considered a prevalent (mis)conception in today's scientific, technological, and control engineering fields. Despite its widespread influence, MR falls short of explaining essential concepts. The relationship between reality and our representational tools remains a puzzling mystery, and MR fails to account for the existence of different yet empirically equivalent descriptions, theories, and models. Moreover, the fixed structure postulated by MR contradicts the notion that the boundary between the objective and subjective may not be as distinctly defined as commonly thought [10].

The various practical design methodologies in control systems design are merely distinct approaches to understanding, interpreting, and engaging with reality. Reality encompasses a complexity beyond the scope of any single perspective. Our unique conceptual frameworks, languages, viewpoints, and biases shape our perceptions, knowledge, and actions. It is impossible to transcend these influences completely and perceive reality in its true form, as it exists independently of our subjective experiences.

MR exceeds its traditional role as an epistemological position in the philosophy of science or prevalent perspective in science and technology. It intersects with *relativism* by suggesting a fixed reality beyond our full comprehension. Putnam highlights that relativism can be seen as a form of MR, although pessimistic. The concept of IR acknowledges varying degrees of realism in different contexts. Relying solely on relativism or instrumentalism is inadequate, as it is crucial to distinguish between genuine and superficial success in prediction and control.

A comprehensive view of realism should recognise that our knowledge is shaped by interpretations, theories, and models influenced by our perspectives and interests while acknowledging that we maintain genuine beliefs about specific objects and properties. Putnam asserts that claims about real existence are tied to broader frameworks or conceptual schemes. Making realistic judgments within these frameworks allows for objective evaluations despite diverse perspectives. While absolute criteria may be lacking, objective standards within conceptual frameworks still exist.

The implication of Putnam's propositions in the context of control system design suggests that the structure chosen by the designer determines the concept of reality within a control system. The language and formal descriptions used by the designer create the framework through which the control system operates. While various alternative formal descriptions may be available for modelling a control system, the key requirement is that the system functions effectively regardless of the specific model used.

Let *objective criteria* refer to standards, principles, or rules that are universally applicable and independent of designers' opinions, biases, or perspectives. In the context of control systems, objective criteria are the rules, constraints, or conditions governing the system's behavior, ensuring its proper functioning and achieving desired outputs. In other words, objective criteria are universal, independent, unambiguous, and consistent. Examples of objective criteria in control systems include:

System specifications: Design requirements
Performance metrics: Measures of the system's effectiveness
Safety standards: Guidelines that define safe operation.

These criteria are a reference point for evaluating system performance, making decisions, and optimising system behaviour. By adhering to these objective criteria, control systems can operate reliably and achieve desired outcomes, even in complex and dynamic environments. In contrast, *subjective criteria* are based on personal opinions, biases, or preferences, which can lead to inconsistent or unreliable decision-making. In control systems, relying on subjective criteria can result in unpredictable behaviour, instability, or poor closed-loop performance.

Despite the diversity of the mathematical and engineering frameworks used to articulate the functioning of control systems for understanding, analysing, and designing effective closed-loop systems, there are objective criteria within these descriptions that govern the behaviour of the control system and ensure its proper functioning. This connection to realism is vital for the successful control of real-world systems, as it is through adherence to objective criteria that the control system can effectively operate and achieve desired outputs.

It can be argued that IR is not a substitute for a stronger form of realism. Rather, it offers a pragmatic understanding of our knowledge and its limitations and may be the only sense in which we can be realistic at all.

Despite the existence of multiple conceptual schemes, IR does not restrict the development of new and better representations, theories, or models. IR allows for integrating previous alternate descriptions into new, more comprehensive frameworks. Alternate descriptions can converge, becoming part of broader theories or models. This convergence can occur through refining existing descriptions, theories, or models or deliberate decisions and choices.

Pluralism is not an obstacle to knowledge but a natural and necessary aspect of scientific inquiry. New situations will always give rise to alternate descriptions, theories, and models, resulting in new pluralism. While this may seem challenging, it is an inherent part of our understanding of the world. In other words, IR provides a framework for understanding our knowledge and its limitations, while acknowledging the complexity and diversity of conceptual schemes. By embracing this pluralism, we can continue to develop and improve our understanding of the world, even as new situations arise.

Finally, in the field of control systems design methodologies, IR implies that embracing pluralism and recognising the diverse conceptual schemes that exist can lead to the developing of new and improved representations, theories, and designs. Instead of viewing multiple perspectives as obstacles to knowledge, they should be seen as essential components of scientific inquiry. By utilising an integrative realism approach, control systems designers can refine existing descriptions and theories while incorporating new alternate descriptions into more comprehensive frameworks. This flexibility allows for continually advancing knowledge and understanding in the face of evolving situations and challenges. In essence, by acknowledging the limitations of our knowledge and embracing diverse perspectives, control systems designers can enhance their ability to create and implement effective solutions for various complex control systems. Control systems scientists and engineers must learn to live in a pluralistic world and overcome the resistance to accepting a pluralistic view of control engineering. In particular, it is important to understand how different approaches can have some realistic value despite their pluralities. For more details on the topics of this section, refer to [10] and the references therein.

9.7 Conclusions

Modernism fundamentally changed the approach to analysing and designing control systems in the last century. The modernistic approach strongly relies on mathematical formalism and optimality in the context of mathematical descriptions of dynamical systems. This school is, therefore, based on determinism in control systems, which involves the tendency to overlook uncertainties and assume exact information and conditions.

On the other hand, postmodernism was an intellectual and cultural movement that emerged in reaction to modernism that gained prominence in the latter half of the twentieth century. It is characterised by a departure from some of the core tenets of modernist thought: *certainties* and *grand narratives*. In the postmodern approach to control systems uncertainty is addressed by approaches that deviate from strict rationalism and determinism in their design philosophy. The postmodern methodologies are proposed to range from fixed structure robust controllers to nonlinear time-varying adaptive controllers and nonlinear time-varying intelligent control with underlying soft computing principles. These postmodern design methodologies build a platform for controlling complex, uncertain, time-varying, and unknown systems with ill-defined mathematical models.

Postmodernism also provides a platform for a pluralistic approach to control systems design to overcome some of the problems associated with model-based control techniques and to enhance closed-loop system stability and performance.

Finally, the internal realism concept and pluralism are discussed in control systems modelling and design, which provides a new perspective on reality and control.

References

1. T. Samad, Postmodernism and intelligent control, in *Proceedings of 12th IEEE International Symposium on Intelligent Control* (IEEE, 1997), pp. 37–42
2. J.M. Prausnitz, Chemical engineering and the postmodern world. Chem. Eng. Res. Des. **79**(7), 777–788 (2001)
3. T.L. Ferris, The interplay of modernism, postmodernism and systems engineering, in *2004 IEEE International Engineering Management Conference (IEEE Cat. No. 04CH37574)*, vol. 3 (IEEE, 2004), pp. 1003–1007
4. F. Jameson, Postmodernism, or the cultural logic of late capitalism, in *Postmodernism* (Routledge, 2016), pp. 62–92
5. M.M. Bălaş, V.E. Bălaş, Postmodernism and control engineering, in *Views on Fuzzy Sets and Systems from Different Perspectives: philosophy and Logic, Criticisms and Applications* (Springer, 2009), pp. 377–391
6. C. Virgil Negoita, Postmodernism, cybernetics and fuzzy set theory. Kybernetes **31**(7/8), 1043–1049 (2002)
7. L.A. Zadeh, The evolution of systems analysis and control: a personal perspective. IEEE Control Syst. Mag. **16**(3), 95–98 (1996)
8. E.A. Jackson, A first look at the second metamorphosis of science. (Santa Fe Institute Santa Fe, NM, 1995)
9. D.M. Clarke, Philosophical papers: vol. 1: realism, rationalism, and scientific method. Philos. Stud. **31**, 480–481 (1986)
10. M. Liz, M. Vázquez, J. Aracil, Knowledge, control, and reality: the need for a pluralistic view in control systems design. IFAC Proc. Vol. **29**(1), 829–834 (1996)

Chapter 10
Future Trends in Control Systems: A Journey Through Time

The illiterate of the future will not be the person who cannot read. It will be the person who does not know how to learn, unlearn, and relearn.
—Alvin Toffler

10.1 Introduction

Control systems play a crucial role in shaping the technological landscape of the future. Indeed, Astrom brilliantly describes control as the *hidden technology* and states in [1], "Automatic control systems are increasingly becoming mission critical in the sense that the system fails if the controller fails." Many control scientists and engineers have looked ahead to the challenges, advancements, and applications of control theory and practice over time, predicting the coming years. Exploring the evolution of control systems from the past to the present and into the future has always been important. Many individuals have shared their visions in numerous papers and books. In this Chapter, building on the insights of leading control scholars and engineers and data from surveys conducted among control engineers in the industry over the past nearly 70 years, we briefly review the reported trends and directions. Several papers and reports have been selected, their perspectives are synthesised, and conclusions are drawn. The Chapter commences with the earliest paper from 1963, which focuses on research directions in control. After this, a 1987 paper addresses challenges in control, as detailed in Sect. 10.3. An in-depth analysis of the impact and future requirements of control theory in the industry is presented in a comprehensive paper based on surveys from the Japanese industry, published in 1999. The results of this paper are presented in Sect. 10.4. Section 10.5 covers essential points from a comprehensive 2003 paper that outlines future directions in control, dynamics, and systems. The impact of control technology is analysed and discussed in two reports published in 2011 and 2104, as the first and second editions, respectively. The results of these reports are briefly presented in Sect. 10.6.

A. Khaki-Sedigh, *Control Systems: A Historical and Philosophical Perspective*,
https://doi.org/10.1007/978-981-96-7426-8_10

Additionally, this Chapter briefly discusses a 2017 review paper on systems and control for the future of humanity in light of advancements in information and communication, presented in Sect. 10.7. Following the 1999 paper on control in the Japanese industry, a significant study on control-industry engagement was conducted in 2020 and is detailed in Sect. 10.8. Section 10.9 provides a brief overview of a 2023 report that presents the results of a comprehensive study on societal drivers, challenges, emerging methodologies, and future issues not previously addressed in control systems analysis and design, encapsulated in the 2030 control systems road map. Finally, the Chapter concludes with a brief discussion of the ethical considerations in control systems engineering in Sect. 10.10.

It is to be noted that the papers and reports discussed in the section were authored by various scholars at different points in time. These papers and reports are chronologically presented, allowing for an understanding of the evolution of ideas and research on the topic over time.

10.2 The Fuller 1963 Paper on Directions on Research in Control

One of the earliest papers written on the future trends in control systems is the 1963 paper by Fuller [2]. It discusses the evolution of control systems, suggesting that they were considered degenerately simple in the past. However, it argues that control is currently undergoing a transformation towards increasing complexity. The development of control theory is critically reviewed, and the conclusion is drawn that the control theory of the 1950s and early 1960s relied on overly simplified mathematical models and is somewhat disconnected from actual contemporary practices, which have been predominantly empirical. The dominant empirical practices in control engineering were primarily focused on the hands-on, experimental approach to understanding and improving the performance of control systems in various industries, such as aerospace, automotive, and manufacturing. During this period, control engineers relied on trial and error, data collection, and analysis to fine-tune control systems. They used techniques like trial-and-error tuning of proportional-integral-derivative (PID) controllers.

On the other hand, theoretical control theory relied on exact linear time-invariant mathematical models that resulted in engineering failures in practice. The issues of noise, disturbance, and time-varying parameters were other challenges highlighted by Fuller and were dealt with in the 1960s and 1970s control theory research. For more details, refer to Chap. 4.

Therefore, Fuller anticipated two general directions for future control research. Firstly, an effort to ground the theory of simple single-variable control systems on a more realistic basis. That is a need for more *accurate mathematical models* that better represent real-world systems. Although the term *uncertainty* is not explicitly mentioned in the paper, its implications can later be traced in the literature on uncertainty mitigation efforts, as discussed in Chap. 4 and Sect. 5.7. Secondly, empirical development is emphasised, particularly in creating controllers for systems that are becoming more complex in the next stages of technological evolution. Admitting that control is inherently a complex and challenging subject, with its development characterised by a gradual accumulation of insights since the Industrial Revolution, it is stated that there was no comprehensive theory for even the simpler control systems. It is argued that theories relied on simplified models of the plant, and there is a noticeable *gap* from the empirical nature of control design practices. Finally, it is foreseen that future developments in control will move towards an attempt to put the theory of single-variable control systems on a realistic basis to show that the conventional 3-term controller (or something like it) is close to optimum and empirical development of controllers for multivariable systems and self-adjusting systems. The multivariable control theory and practice flourished from the late 1960s and 1970s. This period witnessed significant advancements and widespread applications of multivariable control theory in various industries, such as aerospace, automotive, chemical, and process control. For the developments in self-adjusting or adaptive control systems, refer to Sect. 4.3. However, as discussed in Sect. 10.4, a 1999 study highlighted that the interaction challenge in control engineering practice remained an ongoing issue of more importance than the problem of time-varying parameters.

10.3 The 1987 Assertion on Challenges to Control

In early 1986, an open call for participation in a workshop to discuss the challenges and future of control systems was published in the April issue of the IEEE Control Systems Magazine, and in June, during the 1986 American Control Conference, a final structure of the programme was finalised. The workshop's findings, which are the contributions of many prominent control systems theorists and engineers, are reported in [3], highlighting the needs and future directions for control and systems theory. The key points from a late 1980s perspective are as follows:

- Control theory needed to evolve to handle incompletely known systems, as real-world models were rarely fully understood or linear. Researchers emphasised online model refinement and new model classes while extending control methods to nonlinear, stochastic, and distributed-parameter systems. Computing innovations enabled adaptive control laws, facilitating online learning and improved performance. This was called an *ambitious* agenda of theoretical tasks that was considered a *realistic* goal.
- Conceptualisation of discrete event systems.
- Cross-disciplinary algorithm tools (stochastic control + simulated annealing; queueing theory + perturbation analysis) advanced network optimisation.
- The use of symbolic computation for further extensions of linear geometric concepts to nonlinear systems.

- Modeling uncertainty became formally linked to the convergence properties of adaptive estimation and control algorithms, marking a significant theoretical advancement.
- Interaction between computers, electronic technology, evolving applications, and advancements in fundamental concepts: significant expansions of current theories are required to meet performance demands, and in certain instances, new theories are essential to understand the characteristics of emerging artificial systems.
- Increased complexity was apparent, and it was advocated that future research provide the methodologies, theories, and algorithms that would allow the control engineer to plan, manage, and control systems of unprecedented complexity.

The challenges of the late 1980s and the future challenges agreed upon by the *settled* control systems community of the 1980s are briefly outlined. The research challenges encompassed diverse areas, including multivariable control system design, discrete event dynamical systems modelling, geometric and asymptotic methodologies for nonlinear systems, and experimental data requirements for distributed architectures. These topics were chosen for their technical significance and future potential, though specific applications like economic system control were deliberately excluded due to resource limitations or disciplinary considerations. Two central themes emerged from these discussions: first, how increasing performance requirements reshaped design methodologies, and second, the transformative role of digital computers and microprocessors, which were becoming integral system components rather than mere computational tools. The topics included:

- The need for multivariable robust, adaptive, and fault-tolerant control was emphasised due to the impracticality of describing the dynamics of many practical systems using exact mathematical models. Advances in VLSI technology enabled the reliable and cost-effective implementation of sophisticated multivariable feedback control algorithms in digital controllers.
- Conservatism was seen as a significant challenge to the control system design.
- Implementation of theoretical results was another major concern.
- Structured uncertainty modelling, incorporating quantitative and qualitative plant knowledge, offered an alternative to overly conservative designs.
- Adaptive control demonstrated value in robustness enhancement through robust adaptive architectures, performance quantification via exogenous excitation, nonparametric approaches, real-time frequency-domain uncertainty characterisation, and fault-tolerant system design.
- Combining concepts of control theory and artificial intelligence for fault-tolerant control systems.
- Research priorities extended to four key domains: stochastic control for random disturbance processes, nonlinear control for highly nonlinear systems requiring new observer and compensation theories, distributed parameter systems with spatio-temporal dynamics, and hybrid systems with continuous and discrete states.

The report concluded by emphasising three critical needs for advancing control engineering: Establishing a common technical language to bridge researchers, educators, and industry; Expanding beyond traditional mathematical foundations to incorporate insights from cybernetics, computer science, robotics, and even biology; Developing new frameworks for system integration and real-time control that account for computational architectures and dynamic performance. These interdisciplinary demands underscore the growing complexity of control systems and the necessity for diverse, specialised expertise to drive future innovations.

10.4 The 1999 Report of Future Needs for Control Theory in Industry: The Japanese Industry Case Study

A 1999 paper presents the future need for control theory in industry, resulting from a control technology survey in Japanese industry[1] [4]. This is one of the earliest published results of the industry engagement of control theory.

The paper can provide a historical perspective on the evolution of control theory in industry and how it has changed over a decade from 1989 to 1999. This information could be valuable for understanding trends and future directions at the end of the last century in control theory for industrial applications. Additionally, as Japan is known for its advanced manufacturing capabilities, insights from this survey could adequately address the specific needs and challenges of control theory in industrial systems, which may also be relevant to other industries.

The paper draws on a report produced by the control technology committee of the Japanese Society of Instrument and Control Engineering (SICE). In 1995, this committee conducted an extensive control technology survey of the Japanese industry by sending inquiry letters, analysing the effective responses, and plant visits.

The research examined three domains: target plants, control technologies, and development processes. Industry data demonstrated a clear evolution from conventional PID to sophisticated methods like LQG/H∞, MPC, fuzzy control, and enhanced PID techniques. This trend underscores the sector's focus on performance optimisation through advanced control solutions. This indicated that industries have transitioned from using conventional PID control to more advanced and sophisticated control techniques to improve efficiency, accuracy, and overall control system performance.

Also, the control strategies were primarily aimed at addressing specific practical challenges, focusing on managing loop interaction in multivariable plants, handling disturbances, addressing dead time, dealing with parameter changes, managing nonlinearity, and addressing instability. In comparison with a study from 1989, the challenges related to interaction, disturbance, and parameter changes have increased in number, which is consistent with the assessment made by Fuller in 1963 in Sect. 10.2. This comparison indicated a shift in the nature and prevalence of control

[1] A similar investigation was done in 1989, and the results were published in Japanese.

problems over time, reflecting the evolving needs and complexities within industrial control systems.

The assessment framework analysed 18 techniques across PID variants (I-PD, decoupling, etc.), modern methods as called by the paper (MPC, H∞, adaptive), and intelligent systems (fuzzy, neural networks), comparing adoption metrics against 1989 benchmarks to reveal technological evolution patterns.

The utilisation of the mentioned techniques is as follows. Advanced PID control was widely adopted, with approximately 30% of respondents implementing this type of control in their facilities. MPC and fuzzy control were the most commonly used among modern and intelligent control techniques, with just under 40% of factories implementing these methods. The application of modern control theory, such as LQG controller, observer, Kalman filter, and H-infinity control/analysis, was in 10% of the surveyed factories, but it was asserted that these applications rapidly increased within a short timeframe.

Compared to the 1989 survey, the application of decoupling PID, dead time compensation, Kalman filter, MPC, H-infinity control/analysis, rule-based control, fuzzy control, and optimisation showed a significant increase. Overall satisfaction with control technology had also risen compared to 1989. Advanced PID has been widely adopted and had high satisfaction ratings. While modern control theory was not as widely applied, except for MPC, 60–70% of respondents were satisfied with the results. Satisfaction with intelligent control increased, while LQG and adaptive control slightly decreased. Dead time compensation, MPC, and fuzzy control received high ratings. The impact of neural networks and H-infinity control indicated increasing satisfaction.

A quantitative assessment examined various control methods' technical feasibility and expected impact. Technical feasibility measured implementation likelihood, while expected impact reflected anticipated effectiveness, offering insights into future trends. A summary indicated that advanced PID control was widely implemented with reasonable effectiveness, high-growth techniques such as MPC, fuzzy, and neural networks showed increasing adoption with high expectations, established methods such as LQG, H∞, and adaptive control had practical use but limited expected impact, emerging methods such as sliding mode, and nonlinear control were not yet practical but held some promise, mature techniques such as PID auto-tuning, and rule-based control were already in use but underperforming expectations and optimisation control was practical to use with high expectations. The key challenges in implementation were categorised as:

- *Primary failure factors*: Inadequate process analysis, poor sensor selection, insufficient disturbance countermeasures.
- *Broader environmental influences*: Process modelling accuracy, control performance evaluation, software productivity, data accessibility (real-time/historical) and HMI design.

10.5 The 2003 Declaration of the Future Directions in Control, Dynamics and Systems

Organised in April 2000, the P*anel on Future Directions in Control, Dynamics, and Systems*, comprising leading control scientists and engineers, developed a strategic vision to:

- Identify emerging challenges and opportunities.
- Enhance the visibility of control research.
- Highlight its relevance to national priorities.
- Outline key trends for sustaining field vitality.

The resulting report [5] provided actionable recommendations for stakeholders across academia, industry, and government. The early twenty-first century's advances in computing, communications, and sensing technologies presented exceptional opportunities for control systems to expand their impact across sectors. While control technologies had already demonstrated four decades of successful implementation in aerospace guidance systems, industrial process control, manufacturing automation, and communication systems, newly developed technologies required innovative approaches. These applications boosted societal productivity and contributed intellectually through rigorous mathematical developments, systems-thinking models, and complex system management. Distributed computation and smart sensing propagation created an environment rich with data and processing capabilities, profoundly impacting military, commercial, and scientific domains. As software and physical systems grew more integrated, control theory became indispensable for ensuring high-performance, reliable, and reconfigurable operations amid uncertainty. A pivotal shift emerged as control systems evolved beyond low-level regulation to higher decision-making layers, enabling greater autonomy and enterprise-wide coordination. This expansion into nontraditional areas, from resource allocation to large-scale scheduling, opened new opportunities to improve efficiency, safety, and reliability across systems.

Developing novel methodologies became imperative to fully realise control theory's potential in emerging applications. Key challenges included controlling hybrid systems combining symbolic and continuous dynamics, a capability beyond 20th-century control theory's scope, especially for large-scale implementations. Networked, distributed environments demanded new formalisms guaranteeing stability and robustness while accounting for computational and communication constraints. As feedback systems expanded into enterprise-level decision-making, advancing autonomous coordination required extending robust control principles to higher-level operations.

The increasing complexity of engineering systems also necessitated automated control synthesis with integrated validation, enabling rapid design, redesign, and deployment cycles from modelling to hardware-in-the-loop verification. Perhaps most critically, future systems needed architectures that maintained functionality despite component failures, achieving reliability through autonomous reconfiguration and graceful performance degradation rather than catastrophic failure.

The application areas were identified around the following main topics:

1. Aerospace and transportation.
2. Communication networks, control of networks, and control over networks.
3. Robotics and intelligent machines, evolution from industrial robots to autonomous systems.
4. Biology and bioengineering, synergy with engineering via modeling/synthesis tools and biological principles inspiring new control designs.
5. Medicine and biomedical research.
6. Chemical industry.
7. Emerging technologies, nanotechnology (electronics, biomaterials), thin-film processing, microsystems, supply chain optimisation.
8. Additional domains, environmental science, electromagnetics, economics/finance, Quantum/molecular systems, energy infrastructure.

These areas encompassed a vast array of applications, showcasing the wide-ranging relevance of control concepts at the start of the twenty-first century. It was finally emphasised that "the next generation of challenges will necessitate a fundamental change in control research and education."

10.6 The Impact of Control Technology Reports 2011, and 2014

Two valuable reports on the impact of control technology present an overview, success stories, and research challenges, published in 2011 as the 1st edition [6] and 2014 [7] as the second edition. The 2014 report is a follow-up of the 2011 report. The reports include flyers highlighting a specific achievement or research opportunity in the field up to the publication time of the report solicited from control engineers and scientists worldwide. The second edition consists of an extended number of 68 such flyers. These flyers are presented in two parts: *Success Stories for Control* and *Challenges for Control Research,* covering traditional, new, and emerging applications and technologies.

Reports demonstrate that control technology's considerable impact is comparable to and possibly even exceeded by its anticipated future influence.

Control technology has emerged as a critical driver of innovation, with growing investments across established and emerging sectors. Beyond enhancing existing products and systems, it is foundational for ambitious advances in biomedicine, renewable energy, and critical infrastructure. The increasing complexity of modern technological systems demands interdisciplinary collaboration, where control principles have repeatedly demonstrated their value. Other fields now recognise control theory for its rigorous systems approach and expertise in dynamics and feedback.

These characteristics have led to an increase in the development of large-scale, safety-critical systems.

Despite these contributions, the report notes that control's full societal impact remains underappreciated, even among practitioners. To address this gap, it documents successful applications across industries, identifies promising research directions, and analyses sector-specific challenges. Rather than proposing theoretical frameworks, the study focuses on demonstrated achievements and unrealised potential, highlighting how control systems technology continues transforming engineering practice while its broader significance awaits fuller recognition. This issue is dealt with in Sects. 10.5 and 10.9.

The role of control in both traditional and emerging application domains, including aerospace, process industries, automotive, robotics, biological systems, and renewable energy and smart grids, is discussed in the 2011 report [6]. The interested reader will find historical references, highlights from successful applications, economic and market information, and current and future challenges in this report.

The report highlights key trends in control technology's evolving role across industries. While early applications primarily served aerospace, manufacturing, and building automation, modern implementations now span diverse sectors. Control solutions have demonstrated remarkable scalability, progressing from single-component regulation to multivariable systems, then to plant-wide integration, and ultimately to complex *systems of systems* at the enterprise level. Successful deployment increasingly requires dual expertise, control theory, and deep domain knowledge, promoting stronger collaboration between control specialists and other fields. This expansion is enabled by hardware advances, particularly in sensor and actuator technologies. Quantifying control's full impact remains complex due to its interdependent nature with other innovations, though available economic metrics consistently reveal substantial societal benefits. Significant specific accomplishments of control systems engineering are listed, and some technical details are provided in both reports. It is stated in [6] "… the significance and variety of these contributions is an indication of how extensive the true footprint of control is!".

Finally, the challenges for control research are discussed in both reports, and these cross-cutting research directions provide a forward-looking view of the control systems. Four topics were highlighted in [6], that were gaining increasing interest, both in the research community and for government investment. These topics are networked decision systems, cyber-physical systems, cognitive control, and systems of systems.

A few shared aspects among these diverse topics are also given as:

- The interdisciplinary and multidisciplinary nature of research.
- The Networks' ubiquity, influencing various topics, led to the rejection of centralised approaches favouring distributed, decentralised, coordinated, and collaborative solutions.
- The high degrees of autonomy and heterogeneity in many active subsystems.

- Addressing diverse challenges across various industries, societies, and governments, they encompass complex areas such as smart grids, intelligent transportation systems, intricate infrastructures, and emergency response teams, showcasing their relevance and applicability to numerous sectors.
- Interactions and integration of control systems with real-time platforms, with humans as users, in other roles, and with other systems.

In [7] some of the grand challenges for control up to 2014 are presented and discussed, emphasising future opportunities beyond this date.

10.7 The 2017 Review Paper on Systems and Control for the Future of Humanity

Building on the influential report "Control in an Information Rich World", briefly reviewed in Sect. 10.5, the 2017 review paper [8] aimed to emphasise and demonstrate that systems and control lay at the heart of Information and Communication Technologies across various application domains. The paper presented a global perspective on how systems and control theory intersect with societal and research challenges. It demonstrated the interchange between five critical societal issues and seven key research and innovation frontiers within the discipline while illustrating the growing cross-fertilisation between these domains. The analysis underscores control engineering's vital role in developing solutions for complex real-world problems through its unique systems-oriented approach and capacity for interdisciplinary collaboration.

A recap of the societal challenges facing control and systems for 2017 and the future are as follows:

Human Welfare and Sustainability:

- Health and medical advancements.
- Energy transition and climate change mitigation.
- Sustainable development practices.
- Water resource management (distribution, quality, conservation).

Economic and Urban Systems:

- Smart city development and urban infrastructure.
- Connected, efficient transportation networks.
- Manufacturing optimisation (smart, lean, IoT-integrated).
- Financial system resilience and smart services.

Complex System Management:

- Dynamics, stability, and optimisation of large-scale systems.
- Cyber-physical system integration (transportation, healthcare, energy).
- IoT and Industry 4.0 implementation (distributed control, data analytics).

The main research and innovation challenges of systems and control have been structured into seven domains summarised as follows:

Challenge 1: Distributed networked control systems. Distributed sensing and control are seen to be utilised in numerous applications, and the underlying theoretical challenges in such systems are briefly presented.

- *Core issues in networked control*: Fundamental limitations of interconnections, bandwidth constraints, noise, time delays, signal distortion, and reality gap between ideal (instantaneous) vs. actual (degraded) communication.
- *Design requirements*: Robust control algorithms tolerant to non-ideal communication protocols, wireless system challenges (unreliability, time-varying links, power limits), control laws and network protocols, security measures, and performance guarantees.
- *Emerging system complexities*: Dynamic topologies in mobile vehicle networks, smart grids, resilience against cascading failures and cyber-physical attacks on wireless nodes.
- *Security imperatives*: Real-time intrusion detection/isolation, self-reconfiguring systems under attack, and secure abstractions for control-network interactions.
- *Interdisciplinary collaboration*: Joint efforts across control theory, telecommunications, information theory, and computer science.

Challenge 2: Data-driven modelling, machine learning, and control. Model-based design and optimisation are the cornerstones for designing, operating, and maintaining complex engineering systems. However, future systems will exhibit unprecedented complexity. These systems must dynamically adapt to environmental changes, learn from experience, and maintain verifiable high performance through active control. Therefore, data-driven approaches are increasingly being used. Also, the synergy between machine learning and control theory advanced in deep learning and reinforcement learning creates new opportunities for control systems. This can yield:

- Enhanced tools for system identification and adaptive control.
- Novel solutions for intelligent autonomous systems.
- Hybrid methodologies combining theoretical rigour with data-driven flexibility.

Challenge 3: Complexity and control. Complex engineering systems are traditionally modelled using nonlinear ordinary/partial differential equations, often incorporating time delays. While nonlinear control theory has addressed broad challenges over the past four decades, emerging applications demand advanced observer and controller designs that account for real-world complexities, uncertainties, delays, sampled data measurements, and input/state constraints. These needs ensure that nonlinear control remains vital across engineering, biomedical, and social sciences. Many adaptive, self-organising systems exhibit complexity through numerous simple components (physical, biological, or computational) interacting as a *system of systems* and evolving connection graphs with stochastic or implicit topologies, network structures often hidden within large-scale empirical/simulation data. The growing capability to engineer highly complex systems with profound societal implications drives new research demands.

Challenge 4: Critical infrastructure systems. Critical infrastructure systems, including energy grids, water networks, telecommunications, transportation, and healthcare, are vital to societal well-being and economic stability. Their safety and security are paramount as these systems become more complex and interconnected. Past failures, whether from natural disasters or cyberattacks, have demonstrated their catastrophic societal and economic impacts. The key challenges are increasing complexity and cybersecurity risks. Also, the existing fault diagnosis and control methods are designed for centralised systems with known models, while modern infrastructures require distributed solutions for uncertain environments.

Challenge 5: Cyber-physical system of systems. The increasing focus on Cyber-Physical Systems of Systems (CPSoS), where cyber and physical components interact to achieve system-wide objectives, presents significant challenges and control engineering opportunities. These spatially distributed networks, including power grids, industrial plants, transportation networks, and water distribution systems, require novel approaches to optimise large-scale physical systems while meeting operational constraints. Key research priorities include addressing distributed autonomy, fault tolerance, adaptive decentralised intelligence, cybersecurity threats, emergent self-organising behaviours, human-system interactions, and rigorous validation methods. Successfully managing these interdependent challenges is essential for developing reliable, efficient CPSoS capable of maintaining critical infrastructure performance amid growing complexity.

Challenge 6: Autonomy, cognition, and control. The evolution of autonomous consumer products transforms complex technical systems, enabling them to perform sophisticated tasks and adapt to environmental changes with minimum human intervention. These cognitive control systems, integrated into vehicles, aircraft, smart homes, and industrial applications, improve safety and performance and facilitate collaboration in human–machine teams. Unlike pre-2017 systems that excelled only in predictable environments, modern autonomous systems demonstrate goal-oriented adaptability, learning from experience to operate effectively in dynamic conditions. They can observe their environment, reason through challenges, and cooperate with humans and other intelligent systems. While advances in computing and algorithms have made high-level autonomy theoretically achievable, practical implementation faces significant challenges. The key challenges include developing reliable monitoring and control frameworks that ensure safety in real-world applications. Although data collection is no longer a bottleneck, extracting actionable insights requires blending data-driven approaches with rigorous model-based analysis. This has prompted a growing interest in machine learning techniques and nonparametric models to handle complex, large-scale datasets, an interdisciplinary effort critical to the future of autonomous systems.

Challenge 7: Cyber-physical and human systems. Some modern control systems increasingly integrate human factors, particularly in safety-critical domains like healthcare and transportation. While automation handles routine operations, some complex systems require human-machine interaction, such as in aircraft and power plants. A key challenge lies in designing system dynamics that align with operator expectations through human factors engineering. Developing verifiable monitoring and control frameworks for collaborative human-cyber-physical systems demands interdisciplinary research combining control theory, computer science, and statistics. The future control systems must optimise task allocation and coordination between humans and machines. For examples of systems and control applications, refer to [8].

10.8 The 2020 Industry Engagement with Control

In 2017, the International Federation of Automatic Control (IFAC) recognised the need to enhance industry participation and impact within its activities. To bridge the theory-practice gap[2] and ensure that the industry had a more significant say in and benefited from what IFAC offered. Consequently, several surveys were carried out, and the results from some of these surveys were reported in [9]. This study is more comprehensive and detailed than the 1999 study presented in Sect. 10.4, it also encompasses a larger sample size of participants and a broader range of application areas.

The 2020 control systems status quo evaluated by [9] indicates that despite the theory-practice gap in control science and engineering, it is incorrect to perceive the discipline as struggling for recognition outside academia or that industry applications lack progress. Indeed, the field is not limited, and progress and innovation are happening beyond traditional boundaries.

The metric of industry investment in control system innovation can measure the industry's involvement in control science and engineering, and the patent portfolio is selected as a quantifiable method.

The significant growth in control-related patents over the years, outpacing the overall patent growth rate in some regions, suggests that the industry is actively investing in and prioritising innovation in control technology. This emphasis on patent protection indicates that the industry recognises the value of advancements in control systems and considers them crucial for their technological progress. Examining a patent database for patents that include the term control system in their titles, abstracts, or claims from January 1, 1998, to December 31, 2017, across various global patent offices shows that the number of control-related patents had significant growth over time. The ratio between the initial and final numbers of these patents was 7.60. In comparison, the total number of patents increased by a factor of 3.17 during the same period, indicating that control-related patents experienced more than double the growth rate of the total patents. This indicates that the industry has recognised the significance of advancing control technology, the outcomes of which are

[2] The theory–practice gap problem is addressed in Sect. 6.4.5.

believed valuable and deserving of protection through patents. However, industry participation in the research community was not as promising as the patent's finding, and the studied metric for this was the involvement of industry-affiliated volunteers in research-oriented control organisations.

Ten key points are derived from the surveys and discussions by the control scientists and engineers involved in the study. These points are restated below in a categorised form. For more details, refer to [9].

A. **Observation and Awareness**

- Advanced control technologies exhibit various impacts across domains, with varying views of their value.
- Limited awareness exists within the control research community regarding the real-world impact of advanced methods.

B. **Implementation Challenges**

- Domain-specific expertise is critical for successful deployment (beyond theoretical rigour).
- Implementation frameworks (e.g., architectures, infrastructures) are highly industry-dependent.
- Advanced control goes beyond feedback loops; it embodies a *systems-oriented mindset* integrating modelling, optimisation, and uncertainty management.

C. **Industry-Academia Dynamics**

- Corporate R&D occasionally bridges academic research and industrial adoption, but gaps persist.
- Curricular misalignment with industrial requirements creates significant barriers to effective technology transfer.

D. **Economic and Resource Allocation**

- Cost reduction drives industrial innovation priorities in control technologies.
- Economic expectations influence industry investment in research.

E. **Broad Relevance**

- Control science drives emerging technologies (e.g., AI, robotics, energy systems), necessitating interdisciplinary collaborations.

10.9 The 2030 Control Systems Road Map

The comprehensive 2023 road map for control systems published by the IEEE Control System Society and authored by a group of eminent control theorists and engineers is an excellent guide to the societal drivers, challenges, emerging methodologies, and other related present and future issues in control systems analysis and design [10]. This section summarises the report's key findings.

Control systems comprise specialised methodologies and principles designed to regulate dynamic systems while achieving target performance specifications, particularly when accounting for inherent system and environmental uncertainties. The development of these theoretical foundations and technical approaches has undergone substantial progression across multiple disciplines. Historically focused on basic feedback mechanisms for individual devices or systems regulation, the field has evolved to address optimisation and decision-making challenges in complex, large-scale systems, including systems-of-systems and infrastructure systems.

Digital transformation has been a universal force, advancing traditional engineering applications while introducing control-theoretic frameworks into non-engineering domains such as economic modelling and social system analysis. Parallel advancements in computational power, distributed sensing, and networked actuation have accelerated the transition toward ubiquitous automation. These collective developments necessitate a reconceptualisation of control systems' societal role, one that recognises their potential to systematically address large-scale, multidimensional challenges ranging from smart city infrastructure to climate-resilient energy grids.

Three key issues in control systems essential for addressing *societal challenges* are *societal drivers*, *technological trends*, and *emerging methodologies*. By integrating societal needs with technological advancements, emerging methodologies are developed to optimise control systems for various applications. These methodologies utilise advanced technologies to enhance control system performance and address evolving societal needs, ultimately driving innovation and progress in control systems. In what follows, these issues are briefly discussed. For more details, refer to [10].

Societal Drivers. Societal drivers refer to the key challenges and issues that influence the development and evolution of control systems design. The societal drivers addressed include climate change, healthcare for quality of life, smart infrastructure systems, the economy of sharing, and resilience in societal-scale systems.

Climate Change Mitigation and Adaptation. Control systems have a foundational capacity to address climate change issues through innovative solutions across infrastructure, energy, transportation, industry, and agriculture. Control methodologies enhance climate resilience and support mitigation efforts by integrating climate variables into critical systems, optimising resource-constrained processes, and exploiting interdisciplinary collaboration.

Healthcare and Quality of Life. Control systems drive advancements in healthcare through systems-level medical technologies, neuro-engineering, and closed-loop medical devices. Applications like artificial pancreas systems, closed-loop anaesthesia, and cardiac-assist devices demonstrate their role in adaptive, data-driven therapies, improving healthcare outcomes and patient quality of life.

Smart Infrastructure Systems. The development of smart, adaptive infrastructure requires interdisciplinary collaboration and validated testbeds. Control systems enable dynamic, integrated infrastructures that respond to environmental and societal demands, spanning energy, transportation, water, and agriculture. Multiscale modelling and behavioural control strategies are critical for resilience in these complex systems.

The Sharing Economy. The sharing economy can be framed as a feedback control problem, where control theory optimises resource allocation among agents. Transactive control[3] and multilevel optimisation ensure fair distribution, stability, and cooperative system performance in peer-to-peer exchange markets.

Resilience of Societal-Scale Systems. Resilient societal infrastructures depend on control systems for network design, risk mitigation, and resource allocation. Control methodologies strengthen critical infrastructure planning and vulnerability management against global challenges by integrating monitoring tools, economic incentives, and emerging technologies.

Technological Trends. Technological trends refer to the advancements and developments in technology that are shaping the evolution of control systems. They are essential for driving innovation, shaping the future of control systems design, and providing new opportunities for enhancing the performance and capabilities of control systems in various domains. The knowledge of these trends and utilising them effectively can lead to advancements in control systems that address emerging challenges and opportunities in the field. The key technological trends identified are the roles of artificial intelligence and big data, large-scale electrification, biological engineering, and robotics.

AI and Big Data. Integrating AI and Big Data into sociotechnical systems presents a crucial area for the control systems field. This convergence enables advanced capabilities in multi-timescale decision-making, simulation-to-reality transfer, and robust autonomous operation, while control theory provides essential mathematical foundations for reliable AI deployment in safety-critical applications. The complementary relationship between machine learning's data-driven modelling and control theory's dynamic optimisation frameworks drives innovations across both fields, particularly for Industry 4.0 implementations.

Electrify Everything. The global electrification movement presents significant opportunities for control systems to coordinate distributed energy resources, optimise building and transportation energy networks, and integrate renewable generation with evolving demand patterns. Sophisticated control strategies are crucial for managing these complex, multi-scale energy systems while maintaining grid stability and policy compliance.

Engineering Biology. Engineering biology applies systematic design principles to biological systems, enabling breakthroughs in health, sustainability, and industrial applications. Control systems theory is critical for overcoming inherent challenges

---

[3] Transactive control is a mechanism used in control theory and economics to manage the exchange of goods or services between different agents in a decentralised system. It involves allowing agents to negotiate and make decisions based on the value they place on the resources they have and the resources they need. This negotiation process can involve setting prices, establishing rules for transactions, and dynamically adjusting allocations based on supply and demand conditions.

in these complex living systems, particularly in managing uncertainty across scales, from molecular interactions to multicellular behaviours.

Robots in the Real World. Achieving effective robotic deployment in unstructured environments requires control systems that simultaneously address operational resilience, safety, and performance. This necessitates an integrated approach spanning the entire control hierarchy, from low-level actuation to high-level cognitive functions incorporating environmental perception, situational awareness, and autonomous decision-making supported by AI frameworks and digital twin technologies. The fundamental challenges involve developing control architectures capable of handling dynamic uncertainties beyond laboratory conditions while maintaining stability guarantees, alongside implementing safety-certified strategies for human-robot interaction. The future will likely see multi-robot systems utilising distributed computation and communication to achieve their goals in complex environments.

Emerging Methodologies. Emerging methodologies are the new approaches, strategies, and techniques in control systems to address the evolving complexity of societal needs and goals. These emerging methodologies are designed to analyse, synthesise, and optimise control systems using advanced technologies and innovative concepts to achieve high-level objectives. The emerging methodologies include Learning and Data-Driven Control, Safety Integration in Autonomous Systems, Resilience of Cyber-Physical Systems (CPS), Human-Centric Control Systems, and Control Architectures.

Learning and Data-Driven Control. Practical applications drive the fruitful integration of machine learning and data-driven control. Revisiting the foundations of systems and control is essential to utilise the synergies between these disciplines, focusing on the intersection of dynamics, scalability, learning, and architecture. A paradigm shift in control over the past decade[4] is the interaction of machine learning and control systems. Learning-based components are vital in processing sensor data, depicting uncertain system behaviour, and aiding decision-making based on context and observed data. Recent advances in machine learning technologies and methodologies bring new opportunities and challenges. A reassessment of the theoretical foundations of learning and data-driven control is essential to utilise these advancements fully. Recent technological advancements have restored interest in data-driven control methodologies driven by improved sensor technology, enhanced computational capabilities, widespread high-performance communication infrastructure, advanced machine learning algorithms, and the growing accessibility of open-source tools and datasets. While these developments create new opportunities in traditional control applications, they simultaneously introduce significant theoretical and practical challenges for implementing data-driven feedback control in novel environments. The field faces a hierarchy of research challenges ranging from immediate technical hurdles to long-term foundational goals. This landscape presents both the transformative potential and fundamental limitations of data-driven approaches in

[4] For the details of paradigm shift to data-driven control from Kuhn's perspective, refer to Sect. 8.4.4.

control systems engineering, requiring continued theoretical development alongside practical implementation studies.

Safety-Critical Systems. Ensuring the safety of intelligent and autonomous systems is fundamental to developing human trust and enabling broad adoption. A rigorous, formal safety representation is essential for synthesising, verifying, and certifying safe behaviour in real-world applications. For example, semi-autonomous aircraft must adhere to strict operational constraints, while robots must navigate complex environments without collisions. Current approaches often rely on ad hoc, domain-specific solutions, which may overlook critical edge cases, leading to catastrophic failures. Consequently, there is a pressing need for systematic, safety-critical control methodologies that provide verifiable guarantees. Such frameworks must support the design of reliable autonomy across diverse applications, including autonomous vehicles, aerospace systems, advanced robotics, and Cyber-Physical Human Systems (CPHS), ensuring robustness against real-world uncertainties while maintaining operational integrity.

Resilient Cyber-Physical Systems. Developing robust cyber-physical systems (CPS) protects vital infrastructure against cyberattacks, extreme weather, and disruptions. Resilient CPS must maintain performance during normal operations while adapting dynamically to emergencies through real-time reconfiguration. Key challenges include improving threat detection, safety-preserving control, and autonomous recovery. Advancements in adaptive modelling, real-time monitoring, and reconfiguration and restoration architectures will enable infrastructure to withstand unpredictable disruptions while ensuring continuous operation.

Cyber-Physical-Human Systems (CPHS). Cyber-physical systems with human integration (CPHS) enable transformative applications in healthcare, manufacturing, infrastructure, and assistive technologies, from medical devices like artificial pancreases to autonomous vehicles and smart cities. Designing effective human-in-the-loop control systems requires advances in safety–critical algorithms, robust adaptation, and computational human modelling to bridge technical and behavioural uncertainties. A significant research challenge in this field is human models' characterisation and adaptability. Including human elements introduces uncertainties in predicting systems, demanding strict safety requirements. In human-centred control design, a primary challenge is developing methods that are safe, resilient to uncertainty, efficient, and capable of learning and adaptation. Reliability, ethics, and liability are crucial in addressing these challenges.

Control architecture. Designing large-scale societal systems integrating sensors, actuators, algorithms, and human interfaces demands rigorous control architecture principles to ensure reliable operation. Unlike conventional hierarchical engineering systems, critical infrastructures like power grids and social networks often evolve through ad hoc, bottom-up development, increasing vulnerability to cascading failures. A unified architectural framework is essential to address these challenges, emphasising multi-timescale layered control for complex, interconnected systems, co-design of sensing, actuation, and communication within each layer, and an integrated, unifying framework for architectural considerations.

10.10 Ethical Considerations

Ethics and social responsibility have not been prominent topics of discussion in the control systems community, and previous reports on the future or directions of control systems design have not independently addressed these issues, leading to a lack of guidelines in ethical control systems design.

The founder of cybernetics, Norbert Wiener, raised the first serious ethical concerns. Refer to Sects. 3.2.3 and 6.2.3 for more details. In the introduction to his book [11], Wiener clearly sees that the second industrial revolution, with the widespread adoption of automation and artificial intelligence, will make most people's skills and labour worthless, as machines would be able to perform tasks more efficiently and accurately. He calls for "a society based on human values other than buying or selling" as a solution. Wiener also writes, "Those of us who have contributed to the new science of cybernetics thus stand in a moral position which is, to say the least, not very comfortable. We have contributed to the initiation of a new science which, as I have said, embraces technical developments with great possibilities for good and for evil. We can only hand it over into the world that exists about us, and this is the world of Belsen and Hiroshima. We do not even have the choice of suppressing these new technical developments. They belong to the age, and the most any of us can do by suppression is to put the development of the subject into the hands of the most irresponsible and most venal of our engineers." Then, he suggests that the best way to address this issue is to educate a large public about the potential implications of their work and to focus scientific efforts on areas that are less likely to be used for military or exploitative purposes, such as physiology and psychology.

Wiener makes it clear that he is aware that the development of cybernetics has the potential to be used to concentrate power in the hands of those who are already unscrupulous. Also, he seemed to be pessimistic about the potential for power to be used for good and instead believed that it would always be concentrated in the hands of those who are most willing to exploit and manipulate others and that advancements in cybernetics would be used to further entrench this concentration of power, rather than to promote a more just and equitable society. The benefits of a better understanding of man and society, which is promised by the development of cybernetics, may not be enough to outweigh the negative consequences of concentrating power. In other words, the good that might come from a better understanding of human behaviour and society may not be enough to counterbalance the harm that is caused by the concentration of power. Wiener concludes his introduction by "I write in 1947, and I am compelled to say that it is a very slight hope."

In [10], ethical considerations are studied from a broad perspective. It is stated that while safeguarding our environment and natural resources, control system designers must prioritise social responsibility by developing technologies that inherently protect human rights, embody ethical principles, and satisfy regulatory requirements, transforming societal values into technical requirements.

In the past, the field of control engineering may not have emphasised ethics much because the technology being developed was not as advanced or pervasive in society. Control systems were primarily used in industrial settings or specific applications where the impact on individuals or society as a whole was limited. Additionally, ethical considerations may not have been as prominent in the overall discourse surrounding technology and engineering practices.

However, as technology advances and becomes more integrated into various aspects of daily life, the potential ethical implications of control systems have become more apparent. Issues such as privacy, fairness, bias, and societal impact are now more widely recognised and discussed in the context of technology development and implementation. This shift in focus reflects a growing awareness of the importance of considering ethical considerations in all engineering aspects, including control systems, to ensure that technology is developed and used responsibly and ethically.

Furthermore, ethics is crucial for control engineering because as technological advancements, such as artificial intelligence and autonomous systems, become more integrated into society, the potential impact on individuals and communities grows.

Control engineers are responsible for designing and implementing systems that have the power to influence various aspects of people's lives, from transportation to healthcare to surveillance. Therefore, considering ethical implications is essential to ensure these systems are developed and utilised responsibly and fairly. By focusing on ethics, control engineers can help promote transparency, accountability, and equity in the design and implementation of technologies, ultimately contributing to a more sustainable and just future for society as a whole.

Before control systems, ethical issues became a great concern for the scientists working in autonomous and intelligent systems. Three pillars were identified that capture the anthropological, political, and technical aspects of ethics and design in these systems. Then, based on these pillars, eight general principles are considered imperative for the ethical design. It is stated that the overall ethical framework can also be applied to a control systems approach to autonomous systems, and these principles should be fulfilled at each stage of problem formulation and technology development. For further details, refer to [10] and the references therein.

10.11 Conclusions

Control systems analysis and design from the past to the present and into the future is a dynamic and ever-changing field that the insights of many leading scholars and engineers have shaped. The exploration of selected papers and reports spanning nearly 70 years presents a progression in understanding control theory and practice, anticipating future trends and advancements in the field. The continuous development and changes in control systems' priorities and design frameworks highlight the importance of adaptability, learning, and innovation to meet the challenges and demands of an increasingly complex technological landscape. Looking towards the future, embracing learning, unlearning, and relearning is essential to remain at the forefront of control systems innovation and development. This Chapter briefly traces the ongoing journey through time in control systems science and engineering. The ability to forecast and adapt to changes is essential for identifying new possibilities and guiding the future of technology. Finally, ethical considerations are important to observe in control systems engineering. These ethical issues are key topics for future research in the philosophical study of upcoming control systems.

References

1. K.J. Åström, Automatic control—The hidden technology, in *Advances in Control: Highlights of ECC'99* (Springer, 1999), pp. 1–28
2. A. Fuller, Directions of research in control. Automatica **1**(4), 289–296 (1963)
3. A. Levis, Challenges to control: a collective view-report of the workshop held at the University of Santa Clara on September 18–19, 1986. IEEE Trans. Autom. Control **32**(4), 275–285 (1987)
4. H. Takatsu, T. Itoh, Future needs for control theory in industry-report of the control technology survey in Japanese industry. IEEE Trans. Control Syst. Technol. **7**(3), 298–305 (1999)
5. R.M. Murray, K.J. Astrom, S.P. Boyd, R.W. Brockett, G. Stein, Future directions in control in an information-rich world. IEEE Control Syst. Mag. **23**(2), 20–33 (2003)
6. T. Samad, A. Annaswamy, *The Impact of Control Technology*, 1st edn. (IEEE Control Systems Society, 2011)
7. T. Samad, A. Annaswamy, *The Impact of Control Technology*, 2nd edn. (IEEE Control Systems Society, 2014)
8. F. Lamnabhi-Lagarrigue et al., Systems & control for the future of humanity, research agenda: current and future roles, impact and grand challenges. Annu. Rev. Control **43**, 1–64 (2017)
9. T. Samad et al., Industry engagement with control research: perspective and messages. Annu. Rev. Control **49**, 1–14 (2020)
10. A.M. Annaswamy, K.H. Johansson, G.J. Pappas,Control for societal-scale challenges: road map 2030 (IEEE Control Systems Society, 2023)
11. N. Wiener, *Cybernetics or Control and Communication in the Animal and the Machine* (MIT press, 1948)

Philosophy Conceptual Index

These philosophical concepts are relevant to the philosophical study of the foundation and application of control systems theory.

Aesthetics

Aesthetics refers to the philosophical study of beauty, taste, and art concerning human experience. Aesthetics in engineering design refers to the intentional creation of visually appealing and harmonious designs that balance form and function.

Aristotle's first principle (Primum movens)

Aristotle's Primum movens is the concept of a first cause or unmoved mover that starts the sequence of events, actions, or changes in the universe without itself being caused by anything else.

Atomism

A philosophical approach that suggests that complex entities can be decomposed into their fundamental, indivisible components, which are atoms. These atoms are distinct, separable, and independent and cannot be further broken down into smaller parts.

Bacon's scientific methodology

It is a scientific methodology developed by Francis Bacon, which emphasises the importance of observation, experimentation, and inductive reasoning in the pursuit of knowledge.

A. Khaki-Sedigh, *Control Systems: A Historical and Philosophical Perspective*,
https://doi.org/10.1007/978-981-96-7426-8

Butterfly effect

The concept of chaos theory suggests that even tiny variations in a system's initial conditions can lead to drastically different outcomes. This phenomenon is exemplified by the butterfly effect, which posits that the slightest perturbation, such as the flapping of a butterfly's wings in Argentina, could potentially trigger a significant event like a tornado in Texas weeks later.

Carnap's confirmation theory

A philosophical framework for evaluating the confirmation of a scientific hypothesis or theory based on the concept of probability. According to Carnap, a hypothesis is confirmed to the extent that it is supported by empirical data and observations. The degree of confirmation is measured by calculating the probability of the hypothesis given the evidence using logical and statistical methods. Carnap's theory emphasises the importance of formal logic and probability theory in evaluating the confirmation of scientific theories and argues that confirmation is a matter of degree rather than an all-or-nothing proposition.

Cause-and-effect relationship

A fundamental concept in philosophy, science, and everyday life refers to the relationship between two events or states where one event (the cause) leads to another event or state (the effect). The cause is typically seen as the antecedent event or situation that precedes the effect, which is the subsequent event or state that follows.

Causal reasoning

A process of logical inference that aims to establish a cause-and-effect relationship between two events or states based on observed patterns and associations. Causal reasoning involves identifying the cause of a phenomenon, event, or outcome by analysing the relationships between variables.

Causation

The relationship between two events or states where one event or state (the cause) directly and reliably produces an effect, where the effect is a necessary consequence of the cause.

Certainty equivalence principle

The principle states that if two events or statements have equivalent causal properties, they can be treated as having equivalent certainty or probability. In other words, if two events or statements have the same causal relationships, they can be considered equally certain or probable. It helps to simplify complex problems under uncertainty by allowing designers and decision-makers to focus on the causal properties of events or statements rather than their uncertainty by acknowledging that causally equivalent events or statements have equivalent certainty or probability.

Circular argument and reasoning

A circular argument is a specific type of argument that is flawed because it relies on the conclusion being argued for as a premise, creating a logical loop. The argument uses the conclusion as one of its premises, making it invalid. Circular reasoning is a broader term that refers to any form of reasoning that relies on itself or repeats itself without providing independent evidence or justification. It can include circular arguments but also refer to other types of flawed reasoning, such as redundancy, tautology, or self-reference.

Circular justifications

It is a reasoning process that relies on itself, where the conclusion is derived from the premise, and the premise is itself derived from the conclusion. This creates a logical loop where the argument appears to be self-contained and self-justifying but lacks independent support or external evidence.

Cognitive

In epistemology, cognition refers to the process of knowing or acquiring knowledge. Philosophers have long been interested in understanding how we form beliefs, perceptions, and concepts and how we use language to represent and communicate knowledge. In the philosophy of mind, cognition refers to the study of mental states and processes, such as perception, attention, memory, reasoning, and decision-making.

Concept of satisficing

Satisficing is the process of choosing in which it is rational to stop seeking alternatives when the agent finds one that is good enough rather than striving for the optimal or perfect solution. This approach acknowledges that decision-makers may not always have the time, resources, or information to find the absolute best solution in situations where uncertainty and complexity are high, and there is limited time to make a decision, and delaying the decision may lead to suboptimal outcomes.

Confirmation theory

It is a field of study in philosophy that focuses on developing and refining models of non-deductive reasoning, enabling logical inference from incomplete or uncertain information. This type of reasoning is commonly used in scientific inquiry, everyday life, and decision-making. Confirmation theory aims to create plausible frameworks that facilitate systematic and rigorous evaluation of evidence, prediction-making, and belief updating.

Corroboration

It refers to the process of confirming or verifying a hypothesis, theory, or claim by gathering additional evidence that supports and reinforces its validity. This involves accumulating further information, data, or observations that are consistent with the original claim, thereby increasing its credibility and confidence in its truth.

Critical rationalism

It is a philosophical approach that emphasises the importance of critical thinking, scepticism, and rational inquiry in understanding the world. Critical Rationalists argue that the only way to achieve genuine knowledge is through a rigorous process of critical inquiry, where all claims are subject to scrutiny and evaluation based on empirical evidence and logical reasoning.

Deductive-nomological view

It is a philosophical approach to scientific explanation and theory construction and attempts to provide a precise and formal account of how scientific theories can be used to explain phenomena. This view posits that scientific explanations consist of deductive arguments, where the conclusion follows necessarily from the premises. Also, the explanations involve nomological statements, which are statements that describe the laws or regularities governing the behaviour of objects or phenomena. The explanations also require initial conditions, which are specific circumstances or states that provide the starting point for the deductive argument.

Emergence

It is a concept with roots in both philosophy and science, and it can be discussed from different perspectives. From a philosophical perspective, emergence refers to the idea that complex systems can exhibit properties or behaviours that cannot be reduced to their individual components or parts. This means that the whole is more than the sum of its parts. Emergence is often associated with concepts like holism, system theory, and non-reductionism.

Empiricism

It is a philosophical approach that emphasises the role of experience, observation, and evidence in understanding the world. In essence, empiricism argues that knowledge comes from sensory experience and data collection rather than through reasoning or intuition alone. Empiricists believe that all knowledge is derived from the senses and that our understanding of the world is shaped by our experiences and observations. They argue that concepts, theories, and laws should be grounded in empirical evidence and testable through experimentation and observation.

Epistemology

It is the branch of philosophy that deals with the nature, sources, and limits of knowledge. It is a vast and complex topic that is concerned with understanding how we acquire knowledge, how we justify our beliefs, and how we distinguish between justified and unjustified beliefs.

Ethics

It is a branch of philosophy that deals with the study of morality and the principles that guide human behaviour. Ethics involves considering the moral implications of actions and determining what is morally justifiable. It concerns understanding the nature of morality, moral values, and moral principles.

Falsification

It is a concept in the philosophy of science that refers to the process of disproving or refuting a scientific theory or hypothesis. It is a key component of the scientific method, as it allows scientists to test and refine their theories through experimentation and observation. Accordingly, a theory is considered scientific if it is falsifiable, meaning it can be tested and potentially disproven.

Holism (Systems-based)

It is a philosophical perspective that emphasises the interconnectedness and interdependence of individual components within a system. Holism asserts that the behaviour, properties, and characteristics of a system cannot be understood by analysing its parts in isolation. Instead, holism views the system as a whole, considering how its components interact and influence each other to produce emergent properties and behaviours that cannot be predicted by examining the individual parts alone.

Hypothetic-deductive model

It is a scientific inquiry method that combines inductive and deductive reasoning to develop and test scientific theories. This approach involves formulating a hypothesis, deducing logical consequences from the hypothesis, and then testing those consequences through experimentation or observation.

Hypothesis

A tentative explanation or assumption made about a phenomenon, which is subject to testing and refinement.

Hypothesis non fingo

It is used to describe Newton's approach to scientific inquiry that avoids making assumptions or hypothesising about the underlying causes of natural phenomena. In essence, Newton believed that scientists should focus on describing the observable phenomena and developing mathematical models that accurately predict the behaviour of those phenomena, without making unwarranted assumptions about the underlying mechanisms or causes. This approach is often contrasted with the hypothetic-deductive method.

Idealisation

It refers to the process of simplifying complex phenomena by abstracting away from irrelevant details and focusing on the essential features that govern the system's behaviour. Idealisation involves making assumptions about a phenomenon, which are not necessarily true in reality, to make it easier to analyse, understand, and predict its behaviour.

Idealism

It is a philosophical perspective that emphasises the importance of abstract concepts, such as ideas, values, and principles, over concrete reality. In science, idealism can refer to the idea that the natural world is ultimately composed of abstract, ideal entities rather than physical particles and forces.

Induction

In science, induction refers to the process of drawing general conclusions or laws from specific observations or experiments. In philosophy, induction is a method of reasoning that involves moving from particular instances to a general conclusion.

Inductive inference

The process of making an inference based on observed data or evidence.

Inductive logic

It is the study of the principles and rules governing inductive reasoning.

Internal realism

It is a philosophical position that combines elements of realism and idealism. It is a metaphysical and epistemological framework that attempts to reconcile the objective nature of reality with the subjective aspects of human experience. Internal realism posits that the world we experience is not independent of our consciousness or perception, but rather, our consciousness or perception plays a crucial role in shaping our understanding of reality. It argues that the external world is not a fixed, objective reality but rather a dynamic, relational process that is influenced by our internal experiences and perceptions.

Deductive logic

Deductive logic is a type of reasoning that aims to establish a high-confidence conclusion based on the truth of its premises. Deductive logic is concerned with deriving a necessary conclusion from a set of statements, where the conclusion follows logically and necessarily from the premises, assuming the premises are true. It is assumed that the premises of a valid deductive argument provide sufficient support for the conclusion.

Logical atomism

A philosophical theory that holds that complex propositions and concepts can be analysed into their most basic, indivisible components, called "logical atoms" or "atomic propositions." These atoms are simple, self-contained units of meaning that cannot be further decomposed or broken down into simpler parts. Logical atomism aims to clarify the structure of knowledge and language by reducing complex ideas to their fundamental, elementary components.

Logical positivist movement

It was a philosophical and scientific movement that emerged in the 1920s and 1930s. It was characterised by a strong emphasis on the power of logic, language, and scientific methodology to understand the world. The movement was primarily associated with a group of philosophers and scientists who were dissatisfied with traditional philosophy and its reliance on metaphysics, which they saw as unclear, ambiguous, and lacking empirical evidence. They believed that philosophy should be grounded in empirical data and concerned with analysing language, logic, and the scientific method.

Materialism

Also known as Physicalism, is a metaphysical and epistemological position that posits that the only fundamental substance in the universe is matter or physical substance. It is a philosophical approach that emphasises the material world as the only reality, rejecting the existence of non-physical entities, such as spirits, souls, or abstract entities.

Mechanism

It is a philosophical and scientific perspective that focuses on the idea that the world is composed of individual, independent components that interact with each other through fixed, deterministic rules or mechanisms. Mechanism postulates that the interactions between its individual parts can fully explain the behaviour of a system without the need for external forces, purposes, or intentional agents.

Natural philosophy

Also known as philosophy of nature, is a branch of philosophy that deals with the study of the natural world and the fundamental principles that govern it.

Objective criteria

Objective criteria refers to standards, rules, or guidelines that are based on evidence, facts, or universally accepted principles and are not influenced by personal opinions, biases, or subjective preferences.

Objectivity

Objectivity refers to the quality of being impartial, unbiased, and detached in one's thoughts, opinions, and actions. Scientific objectivity is the idea that scientific claims, methods, results, and scientists themselves are intended to be free from personal biases, values, perspectives, community bias, and personal interests. Objectivity is considered an ideal for scientific inquiry, a key reason why we value scientific knowledge, and the basis of the authority of science in society.

Plato's metaphor

Also known as the Allegory of the Cave, is a philosophical allegory presented in Book VII of Plato's Republic. It's a powerful and enduring metaphor that has been interpreted in various ways throughout history. Plato uses this metaphor to describe the human condition and our relationship with reality.

Popper's three-worlds ontological framework

Karl Popper's three-worlds ontological framework is a philosophical theory that proposes that there are three separate worlds or domains of existence. This framework is a fundamental aspect of Popper's philosophy, and it has significantly influenced the development of various fields, including epistemology, metaphysics, and philosophy of science.

Occam's Razor

A common philosophical position, often referred to as the principle of simplicity or the simplicity principle, is that all things being equal, simpler theories or explanations are generally preferred over more complex ones. Occam's Razor can be seen as applying the theory of simplicity. It is a philosophical principle that states: "Entities should not be multiplied beyond necessity."

Ontology

It is the philosophical study that examines the nature of existence, being, and reality. It probes deep questions concerning what exists, how things come into existence, and what constitutes reality. Exploring ontology can enhance our understanding of the world and be applied to diverse disciplines, from philosophy to science and technology.

Reductionism

It is a philosophical approach that aims to understand complex systems or phenomena by breaking them down into their constituent parts, studying these parts in isolation, and then attempting to reconstruct the whole from the parts.

Relativism

Relativism in contemporary philosophy means that there is no ready consensus on any one definition. It is a metaphysical, epistemological, and ethical position that holds that knowledge, truth, and moral values are relative to the individual or cultural context in which they are considered. In other words, relativism asserts that there is no objective truth or absolute standard for evaluating knowledge, truth, or morality. Relativism is often contrasted with absolutism, where there is an objective truth or standard that exists independently of human opinion or perspective, and objectivism, where there is an objective reality that exists independently of human perception or interpretation.

Paradigm shift

It is a fundamental change in how we think, understand, or approach a particular subject, field, or discipline. It is a sudden, profound, and often revolutionary transformation in one's perspective, theories, or assumptions that challenges the existing framework and require a significant adjustment in how we perceive and interact with the world.

Phenomenological

It is a philosophical approach that focuses on studying conscious experience or perception. It is concerned with understanding how people experience and interpret the world without making assumptions or judgments about the external world.

Pluralism

It is a philosophical concept that refers to the coexistence and acceptance of multiple perspectives, theories, or approaches in a particular domain or field. It is the idea that multiple, often conflicting, views or frameworks can be valid and valuable in their own right.

Positivism

It is a philosophical and scientific approach that emphasises the use of empirical evidence and scientific methods to study and understand the world. It is based on the idea that knowledge can only be acquired through observation and measurement of objective, measurable phenomena.

Prehension

It is a philosophical concept that refers to the process of grasping or taking hold of an idea, concept, or phenomenon through perception, understanding, or experience.

Premise

It is a statement or assumption that serves as the foundation or starting point for an argument, reasoning, or investigation. It is a proposition that is taken as true or accepted as a basis for further consideration, inquiry, or exploration.

Principle of parsimony *see Occam's Razor*

Process philosophy

It is a philosophical perspective that emphasises the dynamic, relational, and temporal nature of reality. It is a reaction against traditional substance-based or static views of reality, which focus on entities, objects, or substances as the fundamental building blocks of existence.

Rationality (human)

It refers to thinking logically, making sound judgments, and acting reasonably and sensibly. It is often associated with using reason, evidence, and critical thinking to arrive at conclusions.

Realism

It is a broad philosophical and intellectual movement that emphasises the importance of reality, truth, and objectivity. Realists believe that the world exists independently of our perceptions, thoughts, or beliefs. That is, reality is considered to be an objective feature of the world, existing regardless of how we experience or interpret it.

Self-awareness

It is the capacity to be aware of one's thoughts, feelings, and experiences. It is the ability to reflect on oneself, including one's own mental state, emotions, and actions.

Theory-laden observations

It is a concept in the philosophy of science, particularly in the philosophy of science and epistemology. It refers to the idea that scientific observations are not neutral or objective but are influenced by the theoretical frameworks and assumptions that scientists bring to their research.

Unification

It refers to the process of combining or integrating different things into a single, coherent framework or structure.

Verificationism

It is a philosophical position that emphasises the importance of empirical verification or falsification to evaluate the truth or validity of a statement, theory, or concept. In essence, verificationists argue that a statement or theory is meaningful only if it can be empirically verified or falsified.

Subject Index

A. Khaki-Sedigh, *Control Systems: A Historical and Philosophical Perspective*,
https://doi.org/10.1007/978-981-96-7426-8

GPSR Compliance
The European Union's (EU) General Product Safety Regulation (GPSR) is a set of rules that requires consumer products to be safe and our obligations to ensure this.

If you have any concerns about our products, you can contact us on

ProductSafety@springernature.com

In case Publisher is established outside the EU, the EU authorized representative is:

Springer Nature Customer Service Center GmbH
Europaplatz 3
69115 Heidelberg, Germany

www.ingramcontent.com/pod-product-compliance
Lightning Source LLC
Chambersburg PA
CBHW051030140725
29557CB00001B/7
* 9 7 8 9 8 1 9 6 7 4 2 5 1 *